D0929587

THE ANALYSIS AND INTERPRETATION OF MULTIVARIATE DATA FOR SOCIAL SCIENTISTS

David J. Bartholomew
Fiona Steele
Irini Moustaki
and
Jane I. Galbraith

CHAPMAN & HALL/CRC

A CRC Press Company
Boca Raton   London   New York   Washington, D.C.

CHAPMAN & HALL/CRC
Texts in Statistical Science Series

Series Editors
C. Chatfield, *University of Bath, UK*
Jim Lindsey, *University of Liège, Belgium*
Martin Tanner, *Northwestern University, USA*
J. Zidek, *University of British Columbia, Canada*

Library of Congress Cataloging-in-Publication Data

Catalog record is available from the Library of Congress

Visit the CRC Press Web site at www.crcpress.com

No claim to original U.S. Government works
International Standard Book Number 1-58488-295-6
Printed in the United States of America 1 2 3 4 5 6 7 8 9 0
Printed on acid-free paper

Contents

Preface

Multivariate analysis is an important tool for the social researcher. This book is based upon a course that has been given to master's level students at the London School of Economics for many years. The course aims to give students with limited mathematical and statistical knowledge a basic understanding of some of the main multivariate methods and the knowledge to carry them out.

The book is mainly concerned with multivariate methods for elucidating the interrelationships between variables which all have the same status. It is not concerned, for example, with causal models such as path analysis, linear structural relations models, multi-level models, or loglinear models where one wishes to predict some variables given the values of others. These are equally important but are very adequately covered in the existing literature. The distinction we are making is essentially that which is made in elementary statistics courses between a correlation and a regression problem. In those terms, this book is about correlation problems.

It is difficult to be precise about the prerequisites for a course of this kind. Perhaps the best way is to list some of the topics that ought to be covered in preparatory courses. These should certainly include: basic descriptive statistics, ideas of sampling and inference (including elementary hypothesis testing and interval estimation), basic analysis of variance and multiple regression, and the representation of categorical variables in two-way contingency tables including measures of association and tests for independence. This is by no means an exhaustive list, but the topics mentioned will serve as markers of the scope of the material which students will need to have covered. Beyond this, one needs a general familiarity with mathematical formulae and their use and interpretation.

The implementation of all multivariate methods requires the use of a computer. Virtually everything covered in this book can be handled on a desktop computer with standard commercial software packages, such as SPSS, or with special software which we have provided on a Web site. A facility for handling such materials is almost universal among students nowadays and, in our experience, poses no significant problems.

Because we make minimal mathematical demands, our presentation relies heavily on numerical examples and on verbal, rather than mathematical, exposition. It is designed to give insight into the purpose and working of the methods. In no sense is this a cookbook. Multivariate analysis is a sophisticated and delicate tool which can best be learnt by working through detailed examples and attempting to interpret the results of the analysis under guidance of an experienced practitioner. This is the pattern that we have followed in our own course and, for that reason, computer practical sessions are an essential part of the course.

The book has a number of distinctive features. One, which we have already mentioned, is that it treats individual topics seriously and in depth in a non-mathematical way. A second is that we have emphasised the unity of the methods and presented them in such an order that practice gained in the early stages will contribute to the understanding of the more difficult methods that come later. There are many cross-linkages between different parts of the book. To gain the full benefit, it should be followed, as a course, from beginning to end. However, in our courses there have always been a number of graduate students undertaking research training who have attended parts of the course in order to learn about one or two particular methods. In writing the book, we have been aware of the need to provide for this kind of usage, but we have not let this deflect us from our prime purpose of giving a sequential and integrated account.

Perhaps the main novelty of the book is the detailed coverage which we have given to latent variable methods. There has been a substantial growth of interest in these models over the past decade or so, and the second half of the book is designed to provide a good grounding in the full range of models which are now available. While factor analysis is one of the oldest multivariate methods and latent class analysis and latent trait analysis have been widely used for many years, their use has often been focused rather narrowly on specific fields — in the case of the latent trait model, educational testing. Their essential unity and wide applicability, thus, has been obscured. We see them as a coherent group of closely related methods which are capable of becoming important tools for use in a wider range of social scientific research.

This book provides only an introduction. Readers who become interested in a particular topic will need to pursue it further by moving on to the specialist literature. In order to provide signposts to up-to-date and comprehensive treatments, we have included a short list of further reading at the end of each chapter. In addition to providing an authoritative and much more broadly based account than we are able to give here, such books also will provide a good point of entry to a much wider literature. There is also a list of references at the end of the book that gives the sources of various research results upon which we have needed to draw. These are included for the benefit of the reader who wishes to delve deeper but it is by no means a comprehensive listing of the relevant research literature.

The course on which this book is based is given in ten one-hour lectures. In addition, each student is required to attend five two-hour practical classes. The latter also serve a tutorial function in that several of the authors and other assistants are present to help on a one-to-one basis. In no year have all the topics in this book been covered, and the selection has varied somewhat from year to year. We are sure that innovative teachers will find other ways of using this material and we hope what we have provided is flexible enough for this to be done.

The full versions of all the data sets discussed in the book, and the software for carrying out latent variable analyses together with instructions for implementing them can be obtained through Chapman and Hall/CRC Web site at

http://www.crcpress.com/us/ElectronicProducts/downandup.asp

In constructing the index for some items (such as technical terms and symbols) we give only one or two defining or illustrative entries, for others (such as data sets and authors' names) we aim to include almost every entry.

Writing the book has been a joint operation. Most of the main text has been drafted by the three first-named authors using their lecture notes as a starting point. The fourth author contributed new material and also provided constructive criticism as the re-writing proceeded.

We are indebted to many others who have assisted us over the years in various ways. We would single out Susannah Brown who, for many years, played a major part in organizing and conducting the practical sessions. During that time, she was largely responsible for assembling the practical materials and for the smooth running of the course. Panagiota Tzamourani, Amani Siyam, and Anastasia Kakou also prepared materials for computer sessions which have been used here. Olafur Gylfason developed the interface for the GENLAT software available on the Web site. Several colleagues, including Nick Allum and Colin Mills, have suggested data sets that we have used as examples. Other colleagues read parts of the book and gave us the benefit of their experience both on its subject matter and on the technical aspects of producing the figures. These include Martin Knott, Rex Galbraith, Albert Satorra, and Colin Chalmers.

David Bartholomew
Fiona Steele
Irini Moustaki
Jane Galbraith

CHAPTER 1

Setting the Scene

The analysis of multivariate data is part of most quantitative social research projects. Although elementary statistics tells us how to analyze data on single variables, much of the interest in social science lies in the interrelationships between many variables. We are often just as interested, for example, in whether people's views on global warming are related to their political views as in their views on either topic in isolation. When we come to consider many such variables simultaneously, the volume of data becomes large and the pattern of possible interrelationships can be very complex indeed. This book gives methods for exploring such interrelationships.

1.1 Structure of the book

Summarization

All of the methods that we shall describe have as one of their objects the summarization of a set of multivariate data. Summarization is concerned with condensing a large mass of data into some simpler form which is more readily understandable. This may be done in a great variety of ways and few of them are peculiar to data analysis. In all our waking lives, we are continually receiving large amounts of information through the senses. The only way we can make sense of that vast amount of data is to extract the salient features. For the most part, this is done unconsciously and we are hardly aware of the complexity of what is taking place in our brains. Sometimes we carry out this summarization by selecting elements which seem significant or representative in some way. In other cases, we may do it by grouping similar things together and treating them as single entities. In multivariate analysis, we are trying to make some of those processes explicit as they relate to quantitative information. The apparatus for doing this mathematically may seem unfamiliar and abstruse, but the ideas are often simple enough. For example, the notion of projecting an image from a high-dimensional space into two dimensions may seem entirely foreign until we remember that this is what happens in photography. A three-dimensional object in the real world is presented as a two-dimensional picture. Something is lost in the process, but the art of photography lies in choosing the angle of the shot to reveal what is judged to be the essence of the object. There are multivariate techniques which are designed to do essentially the same kind of thing. Because we rely so much on visual imagery, most of the methods treated here are designed to express their outputs in a form which could be represented in one or two dimensions.

Many of the methods which we shall use have close parallels in everyday experience even though we may meet them in an unfamiliar guise. In explaining the methods, we shall make frequent use of such familiar analogies.

However, it may be simpler to begin with the similarities with univariate statistics where the ideas of summarization are much more familiar. The concepts of an average and of standard deviation are well known to anyone who has ever taken a statistics course. These numbers are capable of telling us two important things about very large collections of numbers. An average gives us a good indication of where the numbers are located along the scale of measurement, but it tells us nothing about how dispersed they are. For that, we need a measure of dispersion such as the standard deviation. Both numbers are summary measures, each revealing an important feature of the data. The two together give an economical summarization of the raw data. When a second variable is introduced, the strength of the linear relationship between the two variables may be measured by the correlation coefficient. In all of these examples, we are extracting salient features of the data in a form which is easier to grasp than would be the case if we merely looked at the large mass of raw data.

Summarization does not have to be in the form of numbers. Pictures or diagrams can serve equally well. Pie charts are very familiar to readers of company reports where they are often used as a visual way of presenting a set of percentages. Such diagrams can be made to convey more information by the judicious use of shading or colour. The aim here is to find a way of summarizing a large amount of data in a form which our eye is able to grasp immediately.

The term "descriptive" is often used for that part of statistics which is concerned with summarizing data. We shall use both terms from time to time but "summarization" more adequately conveys the idea that a substantial reduction of the volume of data is involved in the analysis. Similarly, people speak about "exploratory" (data) analysis referring to the fact that they are looking for the main message which the data have to convey without any prior conceptions about what they expect to find.

The first part of the book, comprising Chapters 2 through 5, covers methods which aim to summarize, describe, and explore multivariate data sets.

Generalization

Generalization goes beyond summarization by aiming to discover something about the process which has generated the data, and hence to discover something about a wider class of data of which the present set is only a sample. This, too, is a familiar everyday happening in that we base our expectations on sample experiences. For example, an employer puts a few questions to a candidate in a job interview and infers from the answers something about how that person will perform on the job. This involves generalization from a very limited set of information to something much bigger. The typical elementary statistics course quickly moves on to inference which is, essentially, another term for generalization. Much statistical investigation is based on samples, and the interest is not then in the sample itself but in the wider population from which it has been drawn. In order to establish the link between population and sample, we have to know how the sample was selected. In experimental work, the data may be gathered in a controlled environment with treatments allocated at random. In a sample survey, the sample members may be selected at random. These methods establish a probabilistic link between sample and population, and so the theory of statistical inference depends on the theory of probability. In order to make an inference from the sample mean to the population mean, for example, we need to know about sampling distributions and the t-distribution in particular. The set of assumptions on which all this depends is often described collectively as the "model", and the methods are described as model-based methods.

In many circumstances, however, there is no formal sampling or allocation process. We simply observe what nature or society puts before us. What we observe is certainly subject to uncertainty but it is now less clear how to describe that uncertainty in precise probabilistic terms. This is the case, for example, where a number of schools are required to take part in a testing programme. Random selection may be impossible because of the need to obtain cooperation, and so we have a self-selected sample of those willing and able to take part. We still want to be able to generalize from this particular group but lack the rigorous basis of sampling theory to do it. We may then treat the sample as if it were truly random in order to get some "feel" for the uncertainties involved. This is a defensible procedure as long as we recognize the tentative basis for our generalizations.

The second part of the book, comprising Chapters 6 through 9, covers model-based methods where the primary aim is to make inferences about processes which have generated the data.

Note

The division we have made between summarization and generalization is not as hard and fast as we have made it appear. The assumptions on which the model-based methods are based are not always well founded. Although we shall use the whole paraphernalia of inference, the samples we shall use are not always random and the populations from which they are supposedly drawn are not always well defined. We shall often, therefore, find ourselves using things like goodness-of-fit tests in descriptive mode as measures of agreement between observed and expected frequencies rather than as formal tests. Conversely, our choice of methods in descriptive analysis and the conclusions we shall draw from them sometimes go beyond what can be strictly justified. Nevertheless, the perspective which this dichotomy gives should help readers to follow the progression of chapters.

1.2 Our limited use of mathematics

Mathematics is an enormously powerful tool which enables a quantitative argument to be conducted with precision and rigour. It is therefore indispensable in developing the theory of multivariate analysis. However, very little mathematics is needed to grasp the main ideas behind the methods or to use them intelligently. There is very little use of mathematics in this book, though sometimes mathematical ideas are expressed in words — less precisely but, we hope, more understandably. There are no mathematical arguments or proofs. There are, however, a number of formulae and equations and a certain amount of mathematical terminology. For those who are ill at ease with any kind of mathematics, we believe that the effort to become familiar with the language will be amply repaid. Later, we shall summarize the essentials that are necessary to obtain full benefit from the text.

Most of the main textbooks on multivariate analysis are heavily dependent on mathematics, and it is worth pausing to consider why we are able to dispense with so much of it. Modern multivariate analysis developed around the middle of the twentieth century as a natural outgrowth of univariate (one variable) and bivariate (two variable) statistics. It was natural and fairly easy to generalize the t-test, simple linear regression, and analysis of variance to deal with many variables rather than one or two. This involved moving from the normal distribution, which was the backbone of elementary statistical inference, to the multivariate normal distribution. The special mathematical properties of this distribution lent themselves to many elegant manipulations which drew upon much advanced mathematics. At the same time, it was very difficult to implement any of these methods because of the very limited computing facilities then available. This tended to encourage mathematical as opposed to applied developments of the subject. With the rather special exception of factor analysis, which originated outside statistics, very little use was made of multivariate analysis by social scientists. Perhaps this was just as well because the assumption of multivariate normality on which most of the early theory was founded is very rarely applicable in social science. In fact, continuous variables are the exception rather than the rule in the social field.

The position has been totally changed with the coming of powerful desktop computers. Whereas once computing struggled to keep up with theory, the reverse is true now with computing often leading the way. Computers not only made it possible to implement existing theory, but also stimulated the development of so-called computer intensive methods, such as those in the early chapters of this book, which involve very little formal theory. This transformation is the main reason it is now possible to give a sound, if somewhat incomplete, account of applied multivariate methods.

Matrices and vectors

Although we shall use no mathematical arguments or proofs, it will be useful to be familiar with some mathematical terminology. It will be convenient to

speak of *matrices* and *vectors* as a shorthand for arrays of numbers. A matrix
is simply a rectangular array of numbers. For example, the following array is
a 3×4 matrix, so called because it has 3 rows and 4 columns:

$$\begin{matrix} 21 & 33 & 17 & 9 \\ 13 & 41 & 12 & 37 \\ 11 & 25 & 6 & 19 \end{matrix}$$

A vector is a matrix with only one row (called a row vector) or with only one
column (called a column vector). Vectors and matrices are often enclosed in
brackets to emphasise that they are being treated as single entities. In practice,
of course, the rows, columns, and elements all have substantive meanings but,
for the moment, these can be left on one side.

Very often we shall wish to speak not of individual matrices, but of families
of matrices. One of the most important matrices we shall meet is the *data
matrix*, and there are things we shall wish to say about all such matrices.
Imagine a social survey in which data are collected by questionnaire. Suppose
there are ten questions addressed to 100 respondents. The results can be set
out as a matrix in which the rows represent respondents and the columns
the answers to questions. The first row will give the answers given by the
first individual, in order from left to right. The second row, the results for
the second individual, and so on. The result is a 100×10 data matrix. Such
data matrices will be the starting point for many of the analyses carried out
in this book. For that reason, it is convenient to have a notation which is
sufficiently flexible to accommodate matrices of all sizes. We do this by using
symbols instead of numbers to indicate the various components of the matrix.
We might use x, for example, to represent any element of the matrix. We
identify which element we are talking about by using a pair of subscripts to
identify the row and column from which it comes. Thus x_{24} is the element
in row 2 and column 4; in the matrix given above, x_{24} has the value 37. We
need a notation capable of representing matrices of any size. In the case of a
data matrix, we shall use n for the number of rows and p for the number of
columns. We cannot write a row of p numbers when p is unspecified but we
can write the first few elements and the last as in the data matrix below; the
columns are dealt with similarly.

$$\begin{pmatrix} x_{11} & x_{12} & \cdots & x_{1p} \\ x_{21} & x_{22} & \cdots & x_{2p} \\ \vdots & & & \vdots \\ x_{n1} & x_{n2} & \cdots & x_{np} \end{pmatrix}$$

We can refer to a matrix by a symbol chosen to be suggestive of the contents.
To distinguish matrices and vectors from other quantities, they will be printed
in bold type. Thus, the above matrix might be denoted by $\mathbf{X}$. To save writing
out the full matrix, we can simply give the typical element in the ith row and
jth column. For example, the above matrix could then be written $\{x_{ij}\}$. This
is only sensible if the dimensions of the matrix are clear from the context.

Three other important matrices we shall meet are *distance, correlation,*
and *covariance* matrices. These are square symmetrical matrices in which the
element in row i and column j is equal to the element in row j and column i.
The upper right triangle is thus a reflection of the lower left triangle and so
it is not essential to write out both parts.

Formulae and equations

The most formidable expressions which the reader will meet are some of the
equations. A typical example might appear as follows:

$$f_i(\mathbf{y}) = \alpha_{i0} + \alpha_{i1}y_1 + \cdots + \alpha_{iq}y_q \qquad (i = 1, \ldots, p).$$

This equation conveys a great amount of information in a very compact form.
It would be much harder to express its meaning in words, and it is therefore
worth spending time to unravel its meaning.

An equation, or formula, like this can be thought of as a recipe for calculating the quantity on the left hand side of the "=" sign from the elements
which appear on the right. The quantity on the left, called f, is referred to as
a *function* because its value depends on a number of other quantities (which
appear on the right). The fact that it has a subscript means that the formula
tells us how to calculate a set of functions, not just one. Looking to the extreme right, we have the range of i which in this case runs from 1 to p. So
it is really a formula for calculating p different functions. The expression $(\mathbf{y})$
after f identifies the variables on which the value of f depends — in this case
on q other variables $y_1, y_2, \ldots, y_q$; the bold face for $\mathbf{y}$ is used because a vector
is a special case of a matrix. The right hand side spells out what the form
of this dependence is. It involves the q y's already mentioned, together with
a collection of constants, the α's. (Because they are constants, they are not
mentioned on the left hand side.) These constants have a pair of subscripts
which indicates that they are elements of a matrix. This matrix has p rows
(because i runs from 1 to p) and $q + 1$ columns (because the second subscript
runs from 0 to q). Given numerical values for the α's and the y's, it is then
a straightforward matter to calculate the f's by taking the α's, in turn, multiplying by the appropriate y and adding the results. All of that is conveyed
by a single line of text! It should be noticed that, in addition to the use of
symbols, the pattern of the layout helps to convey the meaning of the equation. A practised reader will immediately recognize the *form* of the equation
and be able to take in its main meaning without paying too much attention
to the detail.

The equation could be made even more compact by the use of the "sigma"
notation which is used to indicate summation (Σ is the Greek "S" for "sum").
Instead of having to write every term in a sum, we then only have to write
one of them. Using this notation, we could write

$$\alpha_{i0} + \alpha_{i1}y_1 + \cdots + \alpha_{iq}y_q = \alpha_{i0} + \sum_{j=1}^{q} \alpha_{ij}y_j.$$

The summation sign (Σ) indicates that we add up all the terms following the sign, obtained by letting j take each of the values in turn between the "limits of summation" indicated by the numbers 1 and q. In this particular case, the gain is very modest and we have preferred to write out the expression in full but there are occasions where the "sigma" notation offers greater advantages.

Many equations that we shall meet are *linear* like the one we have used as an illustration in this section. Others, like the following, from Chapter 8 (equation 8.6),

$$\text{Corr}(x_i^*, y_j) = \frac{\alpha_{ij}^*}{\sqrt{\sum_{j=1}^{q} \alpha_{ij}^{*2} + \sigma_i^2}} \, ,$$

look more complicated but, if broken down into their basic elements, involve the same kind of operation.

Much of the apparent complexity of mathematical expressions arises from a statistician's propensity to add embellishments like *, ˆ or ~ to symbols already encumbered with subscripts and superscripts. The formula immediately above is a good example. This practice is actually intended to make things simpler! Having learnt, for example, that α represents an element of a particular matrix, the addition of a pair of subscripts will tell us which element, a circumflex accent (ˆ) will add the information that it is an estimate we are dealing with, the asterisk that the original α has been standardized — and so on. All of this information is conveyed while preserving the family relationship with the original α.

In the second half of the book, we shall occasionally need to say that some variable has a normal distribution and that it has a particular mean and variance. The conventional shorthand for this is $x \sim N(m, v)$ which is read as "x is normally distributed with mean m and variance v'. In latent variable modelling, we often wish to speak of the distribution of one variable, y say, *given* the value of another variable x. We do this by writing that $y|x$ has such and such a distribution.

Before leaving the topic of mathematics, it is important to emphasise that "non-mathematical" does not mean the same as "easy" in this context. The fundamental ideas remain the same whether we express them mathematically or in words.

1.3 Variables

The numbers which we observe and record in a data matrix may have very different meanings and this fundamentally affects the inferences we can draw from them. They may be measurements of continuous variables such as length or time and, if so, it is meaningful to calculate summary measures such as means, standard deviations and correlations. On the other hand, they may be codes indicating, for example, whether someone answered "yes" or "no" to a question. In the latter case, it is immaterial whether we assign "1" and "0" to the two possibilities or "25" and "7" because the only purpose of the codes is to distinguish one response from the other. Nothing we do subsequently with the codes should depend on this arbitrary feature. Issues of this kind are often discussed in the social science literature under the heading "levels of measurement".

The main distinction required in this book is between *metrical* and *categorical* variables. Metrical variables are those which can be recorded on some kind of scale, like response times, lengths or examination scores, where the interval between two values on the scale has a quantitative interpretation. This makes it legitimate to calculate such things as correlations between pairs of metrical variables and to assume that they have, at least approximately, continuous probability distributions. It is common in statistics to distinguish between continuous and discrete variables. The former can take any value in an interval, the latter only particular values (usually the positive integers). This distinction has no real significance for our purposes so we have chosen to use the term "metrical" to cover both.

Categorical variables arise when individuals are allocated to categories. Country of birth is a categorical variable as is highest level of educational qualification. We can use numbers to code the categories into which individuals fall but those numbers are nothing more than codes. It would be meaningless, for example, to calculate correlations from them. It is not unusual to find this advice ignored and sometimes in ways which are not immediately obvious. For example, if respondents are asked to say whether they "strongly agree", "agree", "disagree" or "strongly disagree" with some proposition it is not unusual to associate equally spaced scores such as 1, 2, 3, and 4 with these categories, thus appearing to turn them into metrical variables. In such cases, the codes are being made to carry more meaning than their arbitrary assignment justifies.

Categorical variables may be *ordered* (*ranked*) or unordered. Highest level of educational qualification is ordered but country of birth is not. (Countries can, of course, be ranked by population, gross domestic product and so forth but, if that were the case, "country" would be merely being used as a crude proxy for a metrical variable.) The expressions of agreement or disagreement mentioned above provide another example of an ordered variable. The numbers 1, 2, 3, and 4 assigned to categories may then be interpreted as the ranks of the categories. The "error" of treating ranks as metrical variables may not be as serious in practice as we have suggested. The correlation between the ranks of two sets of variables is often quite close to the product moment correlation between the variable values underlying the ranks. In such cases, the use of rank correlations may be thought of as an approximation to the proper analysis.

The traditional classification of levels of measurement is as follows: nominal, ordinal, interval and ratio. A nominal variable is the same as an unordered categorical variable. An ordinal variable ranks all individuals, possibly with ties.

Interval and ratio level variables are both special cases of what we have called metrical variables, and it will rarely be necessary to distinguish between them. Ratio level variables have a natural zero point and are necessarily non-negative. Amounts of money, length and weight are familiar examples. The units in which they are measured are arbitrary — money may be measured in dinars or dollars, for example. Conversion factors are available to convert from dinars to dollars, or feet to metres for that matter. They are termed ratio level variables because ratios are independent of the units of measurement. Thus, for example, a camera which costs three times as much as another when priced in dollars will still be three times as costly if the ratio is expressed in any other currency.

Interval level variables also have an arbitrary scale but, in addition, have an arbitrary origin. Many scales constructed for educational or social purposes are of this kind. Measures of clinical depression or social need are usually constructed by summing scores assigned on the basis of psychological tests or demographic characteristics. There is usually no natural origin or unit of measurement for such scales and so we may choose them to suit our convenience. The usual example given for an interval level measure is temperature where the Celsius and Fahrenheit scales, for example, have different units and origins. (Strictly speaking, temperature is a ratio level variable because there is an absolute zero but it is so far from the normal temperatures of meteorology that the fact can be ignored for practical purposes.)

The arbitrariness of the scale and location of many social variables is closely linked to the process of *standardization*. If there is no natural origin or scale, then the results of our analysis should not depend on what values we choose for these quantities. Otherwise, we are adding something to the data which is irrelevant. We are therefore at liberty to choose whatever is most convenient knowing that it will not affect the answer. For many purposes, the best choice is to choose the origin at the mean and to use the standard deviation as the unit of measurement. This is achieved for each column of the data matrix by subtracting the column mean from each element and dividing the result by the standard deviation. For present purposes, the important point to remember is that correlation coefficients depend neither on the scale nor the origin of the variables and, therefore, the fact that one or the other of these are often arbitrary has no effect on the subsequent analysis.

In the second half of the book, where we shall be dealing with factor analysis and related methods, we shall meet an entirely different classification of variables. This is concerned with whether or not we can observe the variable. Variables are called *manifest* or *observable* if they can be directly observed and *latent* or *hidden* if they cannot. Sometimes a variable can only be partially observed as, for example, when we can only observe whether or not it exceeds some threshold. Such a variable may be described as *incompletely observed*. The term *underlying variable* is also used but as this is also used as a synonym for *latent variable*, there is a risk of confusion.

1.4 The geometry of multivariate analysis

One of the principal aims of this book is to express the patterns in multivariate data on two-dimensional diagrams. In that sense, therefore, there is a strong geometrical element in our approach. But geometrical ideas go much deeper than this and that fact is reflected in the language of the subject. Since we shall wish to use that language, it is helpful to know something about the geometry of multivariate analysis even if we are going to make no direct use of n-dimensional geometry.

At the most elementary level, we learn how the ideas of simple correlation can be expressed geometrically. The scatter diagram is a two-dimensional representation of pairs of numbers. An example will be found in Figure 2.1. Such a diagram is a representation of an $n \times 2$ data matrix. The "rows" of the matrix are treated as coordinates and become the "points" of the scatter plot. The form of the relationship between the two variables can be "seen" by looking at the diagram in a way which is more immediate than what we learn from inspecting the two columns of the data matrix. It is natural therefore to speak of the rows of the table as points and the relationship between the variables as linear, curvilinear or whatever it turns out to be. Similarly, we speak of the two-dimensional "space" in which the points are located. The same terminology can be used for a $n \times 3$ data matrix though it is not quite so easy to "see" what is going on in a three-dimensional space. However, when we move on to four or more dimensions, our ability to visualize the points fails. Nevertheless, it is still convenient to continue using geometrical terminology. So we continue to call the rows of the data matrix points and refer to distances between points just as if they were distances we could visualize. In some chapters, especially 5 and 6, we shall carry this use of geometrical terminology further. There we shall need to think of rotation of the axes and projections from a higher to a lower dimension. If we wished to develop the theory of these manipulations, it would be perfectly possible to do so algebraically without any reference to any geometrical notions. All that we shall need here is some intuitive insight into what the theory is trying to achieve and, for this purpose, elementary geometrical terminology enables us to visualize what is going on in simple problems with two or three dimensions and to take on trust that something analogous is happening in higher dimensions.

1.5 Use of examples

Examples play a central role in our exposition and they serve several distinct purposes.

i) Simple examples are used to explain and illustrate the methods. These are a partial substitute for the theoretical treatment which would be given in a more mathematical text. Each step in the argument is executed on an example so that the reader can learn by seeing in detail what is being done. These examples are usually unrealistically small in order to prevent the essentials of the argument from becoming lost in a mass of computation. To that extent, therefore, they can give the impression that the usefulness of the technique is limited since it seems to do little more than reveal what was obvious at the outset.

ii) Once the method has been explained, we shall usually work through one or more substantial examples chosen to illustrate the main stages in the analysis. At this stage, the emphasis changes from how the analysis is done to how it is to be interpreted. This is not easy to do in a textbook because interpretation depends on a thorough knowledge of the context of a study and the purpose for which it is being carried out. Most of the "real" examples we shall use were parts of larger research programmes and something is inevitably lost by removing them from their context. We have attempted to minimize these drawbacks by choosing examples which do not require too much detailed knowledge of the background, and which are small enough to deal with adequately without making them trivial.

iii) A special feature of the book is the section, in each chapter, called "Further examples and suggestions for further work". These sections contain several varied examples. Sometimes they focus on particular uses of the method and sometimes they give a more comprehensive treatment. These examples are intended to extend the reader's experience of the use of the technique by seeing them at work in a variety of situations. It frequently happens that the same data set can be analyzed by several techniques, each one illuminating a different aspect of the problem. It is possible, and desirable, for the reader to go beyond the particular analyses which we have chosen to present. The availability of the data sets either in the book or on the Web site will enable the reader to explore the many options which the various software packages offer. The "suggestions for further work" part of the section title is intended to take the reader beyond the conventional idea of "exercises", which often appear at the end of chapters, into more of a research mode where new ground can be explored. The first step should be to try to reproduce the results we have given in the text. This will enable you to become familiar with the software you are using and, in particular, to find your way around the output. The layout of results has not been linked to any particular software package because these constantly change. In our course, we have used successive versions of SPSS as well as our own software but some examples have been re-worked in S-plus for this book. The next step should be to try variations on our analyses, some of which are suggested in the text, but others can easily be invented. For example, trying other techniques, omitting one or more variables or, where possible, drawing subsamples will provide additional practice and insight.

We particularly wish to counter the idea, common among students, that there is one and only one "correct" technique for any given problem. The various methods presented in this book are better seen as ways of revealing different aspects of the data in a way which may go some way to answering substantive research questions.

A word is in order about our choice of examples. We have tried to avoid overuse of what might be called "textbook" examples. That is, examples which have been selected or tailored to show the techniques up to their best advantage. In real research, such examples are rare and the student brought up on them is likely to think that there is something wrong if the technique does not "work" properly. We think that a method can be a useful research tool even if it leaves much unexplained, or if the model does not fit very well. There are occasions, especially under (i) above, where we need simple examples to illustrate particular points of a method unencumbered by complications. In such cases, we have not hesitated to construct artificial examples or borrow well worn examples from the literature. But when we come to the "Further examples", we have used only real data sets. These require minimal background knowledge to appreciate the motive for carrying out the analysis. We hope that you will find them interesting in their own right and find that your analyses reveal things which were not evident at the outset.

1.6 Data inspection, transformations, and missing data

This is an introductory text and we have not thought it desirable to burden the reader with many of the practical issues on which good applied research depends. Only when one has a clear idea of where one is going is it possible to know the important questions which arise under this head.

It is rarely possible to analyze a data set in the form in which it is collected. There may be obvious copying errors or omissions. Recoding of some categorical variables may have to be done. A few summary calculations of means, proportions and variances and the plotting of histograms will help one to get a "feel" for the data and to identify potential problems. If one is proposing to use a technique such as factor analysis which involves assumptions about the form of the distribution of the observed variables, it is desirable to check that they hold approximately. If not, it may be possible to transform the raw data, perhaps by taking logarithms or the square root, to make them more nearly satisfied. All of this we shall take as read, though you should look carefully at any data before starting an analysis.

Undoubtedly, one of the most common problems at this stage concerns missing data. In practice, some of the elements in the data matrix will be missing for one reason or another. In social research, especially, there are many reasons why this happens. Individuals may refuse to answer some questions or their recorded answers may be obviously "wrong". Candidates in an examination may not have time to answer all the questions. Data may be lost or forgotten. It may, therefore, not be possible to apply the basic operations which we describe without some preliminary adjustment of the data matrix.

The most serious consequence of missing data is that the results of the study may be biased. People who refuse to answer particular questions, for example, may be the very people who hold extreme views on the matter at issue. Dealing with potential biases of this kind is not a straightforward matter, especially if there is a large amount of missing data. The problems are more acute in those techniques which are model-based, and so depend on assumptions about the nature of the sample. The reader who encounters this situation must recognize that the problem needs special treatment and cannot be ignored.

If there are very few missing values, or if they are located in one particular part of the data matrix, it may be possible to proceed without serious risk of bias. For example, if most of the missing values are in one column of the data matrix (that is, they relate to only one variable), that column may be deleted. Something will be lost by excluding what may be a key variable, but it should still be possible to discover what the other variables are telling us. Similarly, if the missing values are concentrated in relatively few rows of the data matrix (meaning that some respondents have given few responses), those rows can be omitted. This is sometimes known as listwise deletion. This too may introduce small biases, but the broad picture should not be seriously affected. Some of the examples used in later chapters have been treated in this way.

One common way of dealing with missing values is by *imputation*, that is, by estimating the missing values in some way judged optimal. When the missing data have been imputed in this way, the completed data matrix can be analyzed as if it had been complete in the first place. We must, however, be more cautious in our interpretation.

For some methods, it is possible to modify the technique to make use of whatever information is available. For example, those methods like cluster analysis and multidimensional scaling, which start from a set of distances between objects, can still function if some of those distances are missing or estimated from incomplete information. Methods which are based on categorical data can sometimes be extended to include "missing" as an extra category. In such ways, the data can be adjusted so that the methods we shall describe can be used.

1.7 A final word

Although this chapter contains introductory material, the relevance of some of it may not be immediately apparent until the reader has some first-hand experience of the methods. We recommend referring back to it as necessary and then, perhaps, reading it again at the end as if it were the last chapter.

1.8 Reading

There are many books on multivariate data analysis which include some of the topics of this book.

Mathematical treatments are given in:

Bartholomew, D. J. and Knott, M. (1999). *Latent Variable Models and Factor Analysis*. 2nd edition. London: Arnold.

Krzanowski, W. J. and Marriott, F. H. C. (1995). *Multivariate Analysis, Part 1: Distributions, Ordination and Inference*. London: Arnold.

Krzanowski, W. J. and Marriott, F. H. C. (1995). *Multivariate Analysis, Part 2: Classification, Covariance Structures, and Repeated Measurements*. London: Arnold.

At an intermediate mathematical level, we would recommend:

Everitt, B. S. and Dunn, G. (2001). *Applied Multivariate Data Analysis.* 2nd edition. London: Arnold.

CHAPTER 2

Cluster Analysis

2.1 Classification in social sciences

Classification is one of the most basic operations in scientific inquiry. It is particularly important in social science, where comprehensive theory is often lacking and the first step in the enquiry is usually to detect some sort of pattern in the data. Methods of classification have long been used in biology, where the grouping of individuals according to species and genus has been the foundation of much subsequent work. Although some early work in cluster analysis was done by biologists seeking to classify plants, much of the stimulus for developing the subject has come from problems in the social sciences, broadly interpreted. The following highly selective list illustrates why we might be interested in finding clusters and what practical purposes they might serve.

i) *Marketing*. Direct mailing is likely to be more effective if it is directed to people with similar characteristics who are likely to respond in the same way. Market segmentation, as it is called, aims to divide the target population into clusters (segments) so that each can be targeted in the manner most likely to achieve a positive response.

ii) *Archaeology*. Artifacts made at about the same time or by the same group of people are likely to be more similar than those originating from different times or peoples. By forming clusters of similar objects, it may be possible to reconstruct something of the history of a region.

iii) *Education*. Schools vary in their performance, and when seeking the reasons for that variation it may be useful to cluster schools so that one can ask what those which appear to be broadly similar have in common.

In this chapter we shall describe and illustrate a number of methods of cluster analysis which are both easy to apply and widely applicable.

The problem which cluster analysis aims to solve is to group individuals in such a way that those allocated to a particular group are, in some sense, close together. It is straightforward to do this if objects are characterised by a single measurable quantity such as income. All that we have to do is to group together those individuals who have similar income. It is true that we shall have to decide what we mean by "similar", but that will be governed by the use to which we intend to put the classification. The problem is more difficult if judgements of similarity are subjective or based upon a large number of characteristics of the objects. For example, when judging the similarity of two schools there will typically be a whole set of possibly relevant characteristics like size, location, ethnic mix, and so on. The question then is how we summarise these diverse bits of information so that we can make defensible judgements of similarity. It is this feature which makes cluster analysis a multivariate technique.

In order to see what is involved in basing distance judgements on more than one variable, consider the case where we have two variables each measured on a continuous scale. If the "objects" were people, we might imagine that we have records giving their ages and incomes and that we want to group them on the basis of those two variables. Suppose we plot individuals as points in the plane. Then their position might appear as in Figure 2.1.

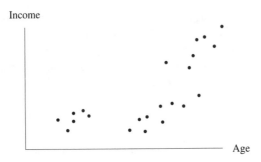

Figure 2.1 *Fictitious data illustrating income and age for several people*

If there were any clustering present, we would immediately recognise it on the figure. There are, in this case, three such clusters which we identify by using the ability of the eye to detect clustering patterns. With three variables, we could imagine points plotted in three dimensions, but beyond that our ability to visualise fails and we need some other way of recognising clusters. Notice, incidentally, that if we had information on just one of these variables — income, say — we would see only two clusters, as the two age groups with low income would be hard to separate. This illustrates how multivariate data analysis can reveal more than the analysis of each variable separately.

In this example, the distance between a pair of individuals is simply defined as their distance apart on the figure. However, the information which we have may not be in the form of measurements on a continuous scale. It may, for example, merely indicate whether or not individuals possess a particular attribute. In such cases, the geometrical representation is not available and some preliminary work has to be done in deciding how we are to measure their distance apart.

Although this example is rudimentary, it serves to identify the two basic steps in any cluster analysis:

i) The measurement of the distance apart of all pairs of objects

ii) The development of a routine, or algorithm, for forming clusters on the basis of those distances

The distances under (i) may be determined either subjectively or by devising a measure of distance based on the observation of a collection of variables. In the former case, it is the human brain which processes the multivariate information available on each object; in the latter, the distance is constructed according to some rational principle.

Before proceeding, we need to clarify the meaning of the term *distance* which we have used to describe how far apart objects are. Sometimes it is more natural to think in terms of closeness or *proximity*. Alternatively, the terms *similarity* and *dissimilarity* are used. The latter have the advantage that they suggest a looser and more subjective assessment of distance which is more appropriate for some of the applications we shall meet. Proximity and similarity stand in inverse relationship to distance and dissimilarity and so measures of one can easily be converted into measures of the other. We shall use the various terms interchangeably but take "distance" as the primary term because it is also central to multidimensional scaling which we meet in Chapter 3.

The first stage of cluster analysis is the construction of distances between pairs of objects. We shall defer the discussion of how this is done until Section 2.4 by which point we shall have a clearer idea of how they are to be used. For the present, we note that the clustering process itself begins with a *distance matrix* which is an array in which the distance between object i and object j appears in the ith row and the jth column. For example, if we have four objects, we have a 4×4 distance matrix:

$$\begin{pmatrix} - & \delta_{12} & \delta_{13} & \delta_{14} \\ \delta_{21} & - & \delta_{23} & \delta_{24} \\ \delta_{31} & \delta_{32} & - & \delta_{34} \\ \delta_{41} & \delta_{42} & \delta_{43} & - \end{pmatrix}$$

where δ_{ij} is the distance between object i and object j. Usually, the distance matrix will be symmetrical, that is $\delta_{21} = \delta_{12}$, $\delta_{31} = \delta_{13}$ and so on. This is because assessments of distance do not usually depend on the order in which we take the two objects. For this reason, it is only necessary to write down half of the δ's — either those in the upper triangle or those in the lower triangle of the matrix. The diagonal may be left blank because these elements play no role in the clustering process. The δ_{ij}'s are sometimes referred to as the observed distances or, simply, as the *observations*.

Methods of cluster analysis may be broadly classified as *hierarchical* or *non-hierarchical*. In a hierarchical method, the clustering process yields a hierarchy in which subsets of clusters at one level are aggregated to form the clusters at the next, higher, level. Hierarchical methods can themselves be divided into *agglomerative* and *divisive* methods. In an agglomerative method, we start by treating each object as a one-member cluster, and then proceed in a series of steps to amalgamate clusters. In such a method, once a pair of individuals has been put together in a cluster, they can never be subsequently separated. This is because any new cluster is formed from clusters already created at previous stages of the process. In a divisive method, we start at the other end, treating the whole set of individuals as a single cluster and then proceed by splitting up existing clusters. Once a pair of individuals have been separated in such a process, they can never come together again. This makes it possible, as we shall see below, to represent the stages in the process by a tree diagram with the branch points indicating where clusters come together or are separated.

In non-hierarchical methods, clusters are formed by adjusting the membership of those clusters existing at any stage in the process by moving individuals in or out. Typically, such methods are more difficult to carry out and they are less commonly used. We shall briefly indicate how such methods might work but shall concentrate mainly on hierarchical agglomerative methods.

In this chapter, we shall treat cluster analysis as a purely descriptive method but the term also covers a wide range of methods, some of which are model-based. We shall meet an example in Chapter 9 on latent class methods where individuals are required to be allocated to categories specified within the framework of a model.

2.2 Some methods of cluster analysis

We shall use a very simple example consisting of only five individuals to illustrate two of the most commonly used agglomerative hierarchical methods, namely the *nearest neighbour* and *farthest neighbour* methods. These are sometimes referred to as *single linkage* and *complete linkage* methods, respectively. Later, we will briefly mention some other methods.

For an example with only five individuals, there is very little that the formal analysis can add to what we can see by inspecting the matrix of distances. We must not therefore expect to learn very much about interpretation from this particular example. Its purpose is solely to define the steps which have to be followed. Imagine that the data concern five customers of a supermarket: Adam, Brian, Carmen, Donna, and Eve, and that information has been obtained from a survey of their tastes and preferences. Suppose also that this information has been analyzed and summarised in the table of distances, Table 2.1. The entries represent how far apart individuals are in regard to their

Table 2.1 *Distance table for illustrating clustering methods*

	A	B	C	D	E
A	–				
B	3	–			
C	8	7	–		
D	11	9	6	–	
E	10	9	7	5	–

potential buying patterns. The smaller the number, the closer the pair. The row and column headings indicate the customer, abbreviated by his or her initial. In a larger scale study with many more customers, the supermarket's objective might be to target its advertising to groups of like-minded customers.

The table has been set out in lower triangular form as the distances are assumed to be symmetrical, and so there is no need to repeat the information in the upper half of the table.

The nearest neighbour (or single linkage) method

This is an agglomerative method in which each customer is initially treated as a separate cluster. First we look for the closest pair of individuals, which involves scanning all the numbers in the table to find the smallest entry. This is the 3 in row B and column A. At the first stage, therefore, Adam and Brian are brought together in a cluster. At the next stage, we construct a new distance table appropriate for the four clusters existing at the end of the first stage. In order to do this, we must specify how distance is to be measured between groups that contain more than one individual. In the nearest neighbour method, the distance between two clusters is defined as the distance between their nearest members. Thus, for example, the distance between the cluster (Adam, Brian) and the individual Carmen is the smaller of the distances from A to C and B to C. That is the smaller of 8 and 7. So in constructing the new distance table, the entry for the row labelled C and the column labelled (A,B) is 7 as shown in Table 2.2.

Table 2.2 *First stage distance table for nearest neighbour clustering*

	(A,B)	C	D	E
(A,B)	–			
C	7	–		
D	9	6	–	
E	9	7	5	–

At the second stage, we simply repeat the procedure for Table 2.2 that we applied in Table 2.1; namely, we look for the smallest value. This is the value 5 which is the distance between Donna and Eve. At the second stage, therefore, we combine D and E into a single cluster. We now have three clusters and require a new distance table giving the distances between these three clusters. This is given in Table 2.3. The only novelty at this stage is finding the distance between the clusters (A,B) and (D,E) which both have more than one member. Adam and Donna are separated by a distance of 11, Adam and Eve are 10 units apart, but Brian is 9 units away from both Donna and Eve. So 9 is the smallest distance and thus goes into the third row and first column of Table 2.3.

The smallest entry in Table 2.3 is 6 indicating that at the next stage we should amalgamate Carmen with the group (Donna, Eve). We then reach a position where there are only two clusters and the only further step possible

Table 2.3 *Second stage distance table for nearest neighbour clustering*

	(A,B)	C	(D,E)
(A,B)	–		
C	7	–	
(D,E)	9	6	–

is to amalgamate them into a single cluster. The foregoing analysis can be conveniently brought together in an *agglomeration table* as follows (Table 2.4).

Table 2.4 *Agglomeration table for nearest neighbour clustering*

Stage	Number of clusters	Clusters	Distance level
Initial	5	(A) (B) (C) (D) (E)	0
1	4	(A,B) (C) (D) (E)	3
2	3	(A,B) (C) (D,E)	5
3	2	(A,B) (C,D,E)	6
4	1	(A,B,C,D,E)	7

What we have achieved by this operation is not quite what we set out to produce. Instead of having arrived at a single set of clusters, we have a hierarchical sequence ranging from the set of five individual clusters with which we started, to the single cluster at which the process ends. We therefore need some way of judging whether any particular set in this sequence has particular claims on our attention. We return to this question after describing a second method.

Farthest neighbour (or complete linkage) method

This is the same as the nearest neighbour method except that the distance between two groups is now defined as the distance between their most remote members. We illustrate the method using the same supermarket example.

The first two stages proceed exactly as in the nearest neighbour method because, up to that point, we were only dealing with clusters containing a single member. The difference arises at the third stage and for this, we need the table of distances between the three groups then existing. This is given in Table 2.5. When judging the distance between the cluster (A,B) and individual C, we now choose the *larger* of the distance between Adam and Carmen and the distance between Brian and Carmen in Table 2.1, namely 8 rather than 7. Similarly, the distance between the cluster (D,E) and individual C is now 7. Finally, the distance between the clusters (A,B) and (D,E) is 11 which is the largest of the four distances in the left-hand bottom corner of Table 2.1.

Table 2.5 *Second stage distance table for farthest neighbour clustering*

	(A,B)	C	(D,E)
(A,B)	–		
C	8	–	
(D,E)	11	7	–

The smallest distance in Table 2.5 is 7. This indicates that at the next stage we should amalgamate Carmen with the cluster (Donna, Eve). The resulting agglomeration table is as follows (Table 2.6).

Table 2.6 *Agglomeration table for farthest neighbour clustering*

Stage	Number of clusters	Clusters	Distance level
Initial	5	(A) (B) (C) (D) (E)	0
1	4	(A,B) (C) (D) (E)	3
2	3	(A,B) (C) (D,E)	5
3	2	(A,B) (C,D,E)	7
4	1	(A,B,C,D,E)	11

It so happens that the sets of clusters produced by the farthest neighbour method coincide with those from the nearest neighbour method, although the distance levels differ at which the clusters merge. Generally the nearest and farthest neighbour methods give different results, sometimes very different, especially when there are many objects or individuals to be clustered. When the two methods (and other methods) give similar results, we may be reassured that the clusters found reflect some true structure in the data.

It is clear that both of the methods considered so far would be very tedious to apply manually to distance tables of a more realistic size. However, each step consists of the very simple operations of selecting the smallest of a set of numbers and constructing the distance table for the next step. This makes it ideal for a computer which can perform very large numbers of such operations accurately and speedily.

A further, important point to notice about both methods is that they only depend on the ordinal properties of the distances. If we were to change all of the distances without changing their order, for example by squaring them, it would make no difference to the clustering. This is a great advantage in many social science applications where the distances are often determined subjectively, and it would not be justifiable to treat them as metrical.

Other agglomerative hierarchical methods

The foregoing methods made rather extreme assumptions about how to measure the distance between two groups. Rather than choosing the nearest or farthest neighbours, it might seem more natural to take distances measured to somewhere near the centre of the cluster. This is what the centroid method does. However, to use the centroid method, we must make stronger assumptions about the distances and the variables on which they are based.

If the distances are Euclidean (that is, straight line distances between the points), we can measure the distances between two clusters by the distance between their averages (centroids). In doing this, we are treating the distances as metrical rather than ordinal. Sometimes the distances may not have been constructed from the coordinates of p continuous variables, but have been arrived at directly and subjectively. Is there then any justification for treating them as if they were "real" distances? To answer this question, we need to remind ourselves that cluster analysis is a descriptive method. Its success or otherwise is to be judged by whether or not it produces "meaningful" clusters rather than the means used for their construction. Obviously, we would prefer a method which reliably produces clustering which can be interpreted sensibly, but this does not rule out *ad hoc* methods of whatever kind. In fact, we shall recommend that several different methods should be used, since well-defined clusters are likely to show up using any reasonable method.

Of the many other agglomerative hierarchical methods available in the various software packages, we mention only one, namely Ward's method. This is of interest partly because it embodies a new general idea about the approach to clustering and partly because, in practice, it often seems to yield the clearest picture of any clustering which is present. At each stage in the clustering process, Ward's method considers all pairs of clusters and asks how much "information" would be lost if that pair were to be amalgamated. The pair chosen is then the one which involves the least loss of information. Information in this case is measured by a sum of squares. If a set of numbers is replaced by their mean, the information loss is defined to be their sum of squares about the mean. The method is designed for the case where the elements of the data matrix are metrical variables since only then is it sensible to compute the sums of squares involved. However, the standard software will produce an answer from the distance table alone by treating them as if they had been computed from continuous variables. The remark we made about the centroid method applies equally here — the usefulness of the method is to be judged by what it produces rather than by the assumptions made along the way.

Other methods

There are a great many other methods available for cluster analysis. Here we merely list some of the possibilities in order to illustrate the rich diversity of methods on offer.

i) *Hierarchical divisive methods.* Here we start with the complete set of objects and, at each stage, divide existing clusters. A particularly simple case arises when the data are binary. One method begins by dividing the objects into two clusters on the basis of a single variable — according to whether or not they possess that particular attribute. The question is then which variable to choose for making the division. This is done by determining which variable is the best proxy for all the variables. One solution is to base this choice on the strength of each variable's association with all the others.

At the second and subsequent steps, one cluster will be subdivided using the best variable for the individuals or objects in that cluster. Different variables might be used in subdividing different clusters.

ii) *Non-hierarchical methods.* These come in many forms. Some use multivariate analysis of variance ideas in the sense that they divide the objects into groups such that the between groups variation is large and the within groups variation is small. One of the most intriguing is known as "Chernoff's faces". The value of each variable determines one feature of a human face — for example, whether the mouth is turned up or down at the corners. The method then hopes to capitalise on our ability to take in facial expressions at a glance and to group faces with similar expressions.

iii) *Model-based methods.* Here we start with the hypothesis that the objects have been randomly sampled from a population made up of several subpopulations. The aim is then to classify the objects according to their subpopulation of origin. Latent class models, which we shall meet in Chapter 9, can be regarded as cluster analysis models and we shall return to this point in that chapter.

2.3 Graphical presentation of results

We now return to the question of how to select a set of clusters from the set of possibilities offered by the foregoing methods. There are two types of diagrams which are frequently used for this purpose, the *dendrogram*, or tree diagram, and the *icicle plot*.

The dendrogram

This has a tree structure with respect to a scale of distances, with individuals, or groups of individuals, located on branches emanating from stems higher up the scale. It can be drawn with the distance scale either horizontal or vertical. Drawn vertically, it resembles an artistic "mobile". It has the property that any two branches hanging from the same stem can be interchanged without affecting the structure (as if the two branches of the mobile were rotated through 180 degrees). However, we will use the horizontal version because it allows easier annotation and labelling of individuals or objects and is consequently often easier to interpret. Again, any two branches meeting at the same stem may be interchanged.

The dendrogram for the nearest neighbour method applied to the data in Table 2.1 is shown in Figure 2.2. There is a horizontal axis showing distance,

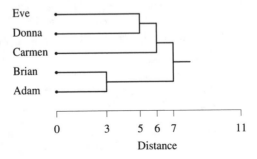

Figure 2.2 *Dendrogram for nearest neighbour (single linkage) clustering*

corresponding to what we called the distance level in Table 2.4. The linking of Adam and Brian takes place at distance 3, which was the distance at which the first amalgamation was made. Donna and Eve come together at distance 5 which was where the second amalgamation was made. Carmen joins Donna and Eve at distance level 6, and finally all five people join at level 7.

Figure 2.3 shows the corresponding dendrogram using the farthest neigh-

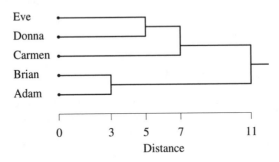

Figure 2.3 *Dendrogram for farthest neighbour (complete linkage) clustering*

bour method, drawn with the same distance scale as in Figure 2.2. The two diagrams are very similar in this case and differ only in the distance levels at which the various clusters join. In general, though, individuals or groups may merge in a different order for the two methods.

The clustering in this example is too ill-defined to give a clear idea how such a dendrogram might help us to choose a set of clusters. We therefore use an artificial example designed to make the point more clearly. The distance matrix in Table 2.7 relates to seven individuals who fall into two distinct clusters. Individuals are labelled numerically here, as is often done in computer programs.

Table 2.7 *Distance table showing a clear pattern of clustering*

	1	2	3	4	5	6	7
1	–						
2	1.5	–					
3	2	2.5	–				
4	10	9	11	–			
5	9	13	12	3.5	–		
6	14	10	11	4	4.75	–	
7	13	12	11	4.5	4.25	3.75	–

It is obvious from inspection of the table, without further analysis, that there are two clear clusters. Individuals 1, 2, and 3 are all close to one another as are 4, 5, 6, and 7. The members of the first group are more tightly packed than members of the second group but the between-group distances are much greater than the within-group distances.

When the results are expressed in the form of a dendrogram for nearest neighbour clustering, we obtain Figure 2.4. The two clusters are clearly identifiable as two groups of branches "hanging" from a single root at the right hand end. It is rare in practice to find anything as clear cut as this but Figure 2.4 gives us a clue about what we should be looking for in general. A tendency for a group of branches to come together at around the same point and then not to be involved in further amalgamations for some distance indicates a possible cluster.

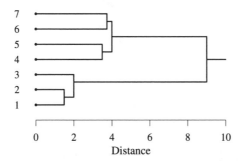

Figure 2.4 *Dendrogram for the data in Table 2.7*

These matters will become clearer when we come to real examples when the clusterings which appear can be related to common characteristics that the members share.

The icicle plot

Whereas the dendrogram can be thought of as a plot of individual against distance, the icicle plot has individual plotted against the stage of merging. Again, these diagrams may be plotted horizontally or vertically, but the name derives from the vertical form which (with some imagination) looks like hanging icicles. We illustrate the vertical icicle plot using the data in Table 2.8. These are extracted from the example on English dialects which we shall use

Table 2.8 *Similarities for a subset from the dialect data*

	V13	V14	V15	V16	V17	V18	V19
V13	—						
V14	63	—					
V15	64	63	—				
V16	54	62	68	—			
V17	72	57	61	59	—		
V18	59	51	56	47	54	—	
V19	38	46	51	42	44	53	—

in Section 2.5. For our present purpose, we merely need to know that the figures in Table 2.8 are *similarities* rather than distances. This means that when looking for the closest pair, we look for the largest number rather than the smallest. Here the individuals are villages labelled V13 to V19.

Carrying out nearest neighbour cluster analysis we obtain the vertical icicle plot given in Figure 2.5.

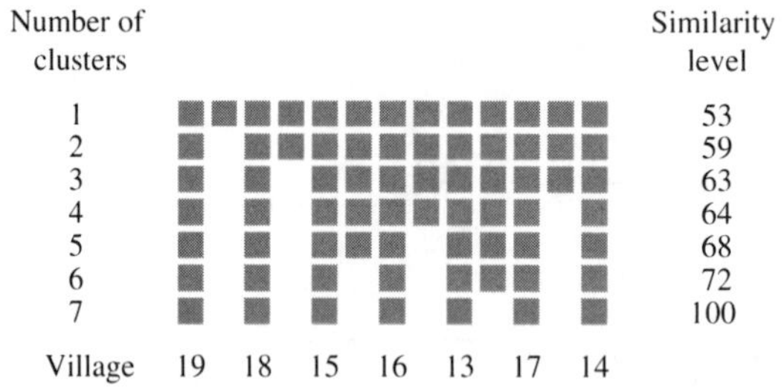

Figure 2.5 *Icicle plot for the data in Table 2.8*

The numbers involved here are too small to demonstrate adequately the usefulness of the icicle plot, and we simply use this example to show how the figure is constructed. The plot enables us to read off the composition of the clusters at each stage of the process. In the diagram, the stage in the clustering process is designated by the number of clusters present. If we look in the first row, when there is only one cluster, we see an unbroken row of shaded boxes extending across the whole table. This is the way the plot shows that all members form a single cluster. In the second row, when there are two clusters, we see a continuous row of shaded boxes extending from village 18 to village 14 with a separate single shaded box under village 19. This means that at this stage of the process, there is one cluster containing the five right hand members and a single cluster consisting of the one member V19. Continuing to move down the table to the row with five clusters we see that there are two 2-member clusters: (villages 15 and 16) and (villages 13 and 17) along with three single-member clusters (village 19), (village 18) and (village 14). The other rows are interpreted similarly. The final row of the table indicates the position when there are seven clusters each consisting of a single village.

To see how a well designed icicle plot can display clusters, refer to Figure 2.8 where all 25 villages are shown grouped into three main clusters, along with two or three stragglers. By adding the distance or similarity level on the right hand side of the icicle plot, we incorporate all the information given in an agglomeration table.

2.4 Derivation of the distance matrix

The first step in cluster analysis is to obtain or construct the distance matrix $\{\delta_{ij}\}$. Broadly speaking, there are two approaches: one that aims to estimate the δ's directly and the other that computes distances or similarities from variables measured on the objects, that is from the data matrix.

In the former case, distance data may be collected directly in an experiment where subjects are asked to give subjective assessments of the similarity (or dissimilarity) between pairs of objects. The colour data, which we shall meet in the next chapter, were obtained by asking subjects to judge differences in colour on a five-point scale. In similar experiments, subjects have been asked to compare pieces of music or the taste of wines. In such experiments, subjects are not usually told what criteria to use to make their similarity judgements.

The other possible starting point is a *data matrix* (see Chapter 1). For each of n individuals, we have values of p variables, giving the $n \times p$ data matrix:

$$
\begin{pmatrix}
x_{11} & x_{12} & \cdots & x_{1p} \\
x_{21} & x_{22} & \cdots & x_{2p} \\
\vdots & & & \vdots \\
x_{n1} & x_{n2} & \cdots & x_{np}
\end{pmatrix}
$$

where x_{ik} is the value of variable k for individual i.

For each pair of individuals, we now ask for some measure of how far apart their respective rows are. The problem is to measure the distance, in some appropriate sense based on the elements of the rows, between any two rows of the table. Note that when distances are constructed from a data matrix, we are implicitly imposing the criteria by which objects are assessed as similar or dissimilar, rather than discovering the criteria through the analysis itself.

Given a data matrix, there are many ways in which the distance (or similarity) between two objects i and j may be defined. We shall review a few of them beginning with the case when all of the x's are continuous.

Distance and similarity measures between objects based on continuous variables

The most commonly used distance measure is Euclidean distance. The Euclidean distance between objects i and j is

$$\delta_{ij} = \sqrt{\sum_{k=1}^{p}(x_{ik} - x_{jk})^2}. \qquad (2.1)$$

For $p = 2$, the Euclidean distance corresponds to the "straight line" distance between the two points (x_{i1}, x_{i2}) and (x_{j1}, x_{j2}).

Often the variables (columns in the data matrix) will be standardized prior to calculating distances. If desired, a weight w_k could be assigned to variable k if it was believed that more importance should be attached to some variables than to others giving:

$$\delta_{ij} = \sqrt{\sum_{k=1}^{p} w_k(x_{ik} - x_{jk})^2}.$$

Another common measure is the so-called "city block" measure, which is the sum of the absolute differences between the pairs of points. Compared to Euclidean distance, this gives less relative weight to large differences. A large difference for a single variable can easily have a dominating effect on the Euclidean distance.

Sometimes the most relevant information for the substantive issues under investigation may be in the comparison of the "shapes" or the profiles of the objects rather than of their "sizes". Similarity need not be judged only by the distance apart of two rows of the data matrix. For example, consider the following examination scores (out of 100) for two students in Mathematics, English, French, and Physics:

Student	Maths	English	French	Physics	Total
1	21	34	17	42	114
2	62	75	58	85	280

Clearly the second student has performed much better than the first in all subjects, so the Euclidean distance between the two students is relatively large. However, although the students differ considerably in their *level* of performance, their profiles have a similar "shape". In fact, the rank orders of their performances in the four subjects are identical, and in this respect they are similar. One way to bring out this feature would be to express each score as a percentage of the total for that student. Thus, the row profile would not show how well a student had done but only the relative contribution of each examination subject. This would only be a sensible thing to do for variables measured on the same ratio scale (here the examinations were all marked on a scale from 0 to 100).

A more general way of eliminating irrelevant differences in size is to scale the scores for each student by subtracting his/her row mean and dividing by his/her row standard deviation. If we then proceed to construct a Euclidean distance between the scaled scores, we would have a measure of how different the shapes of the original profiles were. Interestingly, as simple algebra will show, such a measure is equivalent to using the "correlation" between the two rows as a measure of similarity. (In standard statistics, the Pearson product moment correlation coefficient is calculated from several individuals or objects to measure the similarity between two variables. Here we refer to the same calculation done on several variables to measure similarity between two individuals or objects.) Put another way, we are arguing that similarity in profile of two objects could be revealed by plotting the variables (using the original scores for the two objects as coordinates) and seeing how close they lie to a straight line. For this example, they are very close and the calculated "correlation" is 0.999 compared with a maximum possible value of 1.

This discussion illustrates the important point that the measure we use to define δ_{ij} will depend on what we wish to regard as similar: in this case, students with similar scores on each subject, or students with similarly shaped profiles.

Distance and similarity measures between objects based on categorical variables

If the x's are binary, a measure of similarity could be based on the proportion of variables on which two objects (or individuals) match. Suppose there are p binary variables, each taking the value 0 or 1 for each of n objects. Suppose there are eight variables $x_1, \ldots, x_8$ which for objects 1 and 2, say, take the following values:

Object	x_1	x_2	x_3	x_4	x_5	x_6	x_7	x_8
1	0	1	1	1	0	1	0	0
2	0	1	1	1	0	0	1	1

There are four possible combinations of values for a single variable: $(1,1)$, $(1,0)$, $(0,1)$ and $(0,0)$, where the first value in the pair is the value of x_k for object 1 and the second value is the value of x_k for object 2. Adding up over all variables gives the following frequency table of the number of times each combination occurs:

Combination of values	Frequency for objects 1 and 2	Frequency for two arbitrary objects
$(1,1)$	3	a
$(1,0)$	1	b
$(0,1)$	2	c
$(0,0)$	2	d
Total	8	p

The final column gives the standard notation for a pair of objects and p binary variables: a is the number of variables for which both objects take the value 1, and so on.

We could use as a measure of similarity the ratio:

$$r = \frac{a+d}{p} \, ,$$

where $p = a+b+c+d$ is the number of variables. This measure varies between 0 (no matches at all) and 1 (matches on all variables). It is not difficult to show that the Euclidean distance between the two objects is given by

$$\delta = \sqrt{p(1-r)} \, ,$$

which is a monotonic decreasing function of r. For example for objects 1 and 2 above, the proportion of matches is $r = 5/8$ and the Euclidean distance is $\sqrt{3} = \sqrt{8-5} = \sqrt{8}\sqrt{1-5/8}$.

However, some care should be taken because 1 and 0 may have different meanings for different variables. Sometimes, we may only wish to consider $(1,1)$ as a match. For example, suppose we have a binary variable for nationality where $x = 1$ if a person is a UK citizen, and $x = 0$ otherwise. In this case, if we are really interested in nationality (rather than just whether UK or not), we cannot tell whether two people with $x = 0$ match on nationality. If we do not wish to consider $(0,0)$ as a match, a suitable measure of similarity is a/p. Another measure is Jaccard's coefficient in which $(0,0)$ responses are excluded from both the numerator and the denominator giving:

$$\frac{a}{a+b+c} \, .$$

For categorical data, where the x's have more than two categories, a full set of dummy variables can be constructed for each x. For example, suppose a variable x represents one of four colours red, blue, green or yellow. Then four binary variables x_1, x_2, x_3 and x_4 can be constructed with $x_1 = 1$ when x is red and $x_1 = 0$ otherwise; $x_2 = 1$ when x is blue and $x_2 = 0$ otherwise; and so on. Jaccard's coefficient applied to the full set of binary variables might then be a suitable measure of similarity.

2.5 Example on English dialects

This example is based on a study carried out at the University of Leeds on English dialects. The particular data set used here is taken from Morgan (1981) and is based on results for 25 villages in the East Midlands, shown on the map in Figure 2.6. The study was concerned with the similarities in

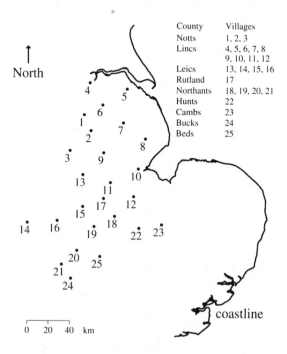

Figure 2.6 *Map of locations of 25 East Midland villages*

dialects among the various villages. A set of 60 items was chosen and members of each village were asked the name which they gave to each item. The data matrix in this case can be thought of as a two-way array with one column for each item and one row for each village. The entry in a typical cell of the table would be the name used for the item heading that column by the villagers represented by the row. The entries in this case are words rather than numbers. If the inhabitants of two villages gave exactly the same names to all items then there would be no difference between villages. The fewer the number of names the villages have in common, the greater the difference between their dialects. The percentage of the items described by the same dialect word in any pair of villagers was used as a measure of similarity.

The complete table of similarities is given in Table 2.9. The 100s on the diagonal represent the maximum similarity (corresponding to agreement on every item) but are not used in the clustering process.

Table 2.9 *Similarity matrix for the English dialect data*

	V1	V2	V3	V4	V5	V6	V7	V8	V9	V10	V11	V12	V13	V14	V15	V16	V17	V18	V19	V20	V21	V22	V23	V24	V25
V1	100	71	58	49	63	64	71	52	46	61	57	39	42	32	32	23	41	39	32	27	28	26	30	36	31
V2	71	100	57	45	63	66	75	56	50	49	60	46	50	34	39	27	47	42	36	36	37	26	33	49	44
V3	58	57	100	48	47	50	52	36	57	52	56	45	53	47	50	42	56	48	43	38	37	30	32	45	40
V4	49	45	48	100	59	53	53	34	33	40	35	30	28	20	19	14	25	24	22	19	20	20	16	26	23
V5	63	63	47	59	100	71	71	60	42	58	53	42	41	27	25	20	38	37	22	22	25	21	25	29	29
V6	64	66	50	53	71	100	68	58	43	61	48	40	36	29	25	22	42	34	24	20	25	28	26	31	32
V7	71	75	52	53	71	68	100	69	43	56	55	47	43	31	36	24	46	36	34	25	31	28	33	41	32
V8	52	56	36	34	60	58	69	100	44	61	48	44	39	23	37	28	36	36	34	25	33	28	32	32	31
V9	46	50	57	33	42	43	43	44	100	52	63	50	48	44	43	36	48	42	29	25	42	41	41	47	47
V10	61	49	52	40	58	61	56	61	52	100	59	53	47	39	41	27	54	43	45	40	37	33	37	32	33
V11	57	60	56	35	53	48	55	48	63	59	100	60	58	43	48	38	54	60	47	40	41	39	37	37	43
V12	39	46	45	30	42	40	47	44	50	53	60	100	48	39	49	35	48	56	38	40	37	55	46	52	43
V13	42	50	53	28	41	36	43	39	48	47	58	48	100	63	63	54	72	59	46	45	48	34	47	46	45
V14	32	34	47	20	27	29	31	23	44	39	43	39	63	100	63	62	57	51	46	49	48	33	46	49	47
V15	32	39	50	19	25	25	36	37	43	41	48	49	63	63	100	68	61	56	51	54	49	40	49	56	53
V16	23	27	42	14	20	22	24	28	36	27	38	35	54	62	68	100	59	47	42	49	47	33	39	49	43
V17	41	47	56	25	38	42	46	36	48	54	54	48	72	57	61	59	100	54	44	42	43	38	46	54	46
V18	39	42	48	24	37	34	36	38	42	43	60	56	59	51	56	47	54	100	53	44	44	40	58	53	53
V19	32	36	43	22	22	24	34	42	29	45	47	38	38	46	51	42	44	53	100	63	58	58	42	63	60
V20	27	36	38	19	22	20	25	29	25	40	40	40	45	49	54	49	42	44	63	100	59	54	44	68	61
V21	28	37	37	20	25	25	31	25	42	37	41	37	46	48	49	47	43	44	58	59	100	47	42	73	62
V22	26	26	30	20	21	28	28	33	41	33	39	55	34	33	40	33	38	40	58	54	47	100	50	51	55
V23	30	33	32	16	25	26	33	28	41	37	37	46	47	46	49	39	46	58	42	44	42	50	100	51	54
V24	36	49	45	26	29	31	41	32	47	32	37	46	57	49	56	49	54	53	63	68	73	51	51	100	72
V25	31	44	40	23	29	32	32	31	47	33	43	45	45	47	53	43	46	53	60	61	62	55	54	72	100

We shall analyze the same set of data in Chapter 3 using multidimensional scaling which, in retrospect, will turn out to be more informative. Nevertheless, as a first step we might want to ask whether we can identify groups, or clusters, of villages which share a common vocabulary.

We begin by applying nearest neighbour cluster analysis to the similarity matrix. Table 2.10 shows the agglomeration table. A different format is used

Table 2.10 *Nearest neighbour agglomeration table, English dialect data*

Stage	Number of clusters	Clusters merged		Distance level	Similarity level
1	24	2	7	25	75
2	23	21	24	27	73
3	22	13	17	28	72
4	21	25	m 2	28	72
5	20	1	m 1	29	71
6	19	5	6	29	71
7	18	m 6	m 5	29	71
8	17	8	m 7	31	69
9	16	15	16	32	68
10	15	20	m 4	32	68
11	14	m 3	m 9	36	64
12	13	9	11	37	63
13	12	14	m 11	37	63
14	11	19	m 10	37	63
15	10	10	m 8	39	61
16	9	m 12	m 15	40	60
17	8	12	m 16	40	60
18	7	18	m 17	40	60
19	6	4	m 18	41	59
20	5	m 13	m 19	41	59
21	4	3	m 20	42	58
22	3	23	m 21	42	58
23	2	22	m 14	42	58
24	1	m 22	m 23	43	57

for the agglomeration table compared with Table 2.4 because of the larger number of villages. Instead of listing all clusters at each stage, columns 3 and 4 of Table 2.10 just show which two clusters are merged. A label prefaced by an "m" denotes the cluster formed at the stage indicated, while a label without an "m" denotes a single village. Thus at stage 1, villages 2 and 7 merge; at stage 2, villages 21 and 24 merge; and at stage 3, villages 13 and 17 merge. At stage 4, the clusters merged are denoted 25 and m 2, which means that village 25 merges with the cluster formed at stage 2 (that is, villages 21 and 24) to form a cluster consisting of villages 21, 24, and 25. One may continue down this table identifying which villages merged at each stage and at what similarity level. The dendrogram (Figure 2.7) shows the information graphically, where it is easier to see patterns of interest.

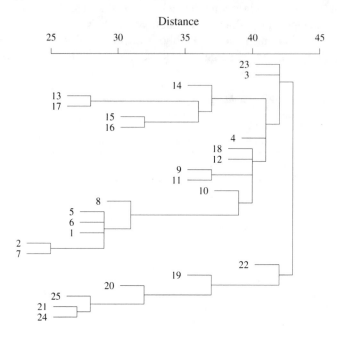

Figure 2.7 *Dendrogram for nearest neighbour (single linkage) cluster analysis for the English dialect data (distance = 100 − similarity)*

The style of dendrogram drawn in Figure 2.7 highlights which villages merged at an early stage (small distance, large similarity level) and which did not merge until a later stage (large distance, small similarity). For example, villages 1, 2, 5, 6, 7, and 8 all merged at a fairly early stage, while 3 and 4 joined them only at a much later stage. Villages 19 to 25 all merged with each other before merging with other villages, though not all at an early stage. Other patterns are also evident from Figure 2.7, which seems to be quite informative.

For completeness, Figure 2.8 shows an icicle plot for this analysis. This shows three clear clusters that together include all but three villages, numbers 23, 3, and 22, which only merge at similarity level 58. The similarity level at each stage of merging is listed on the right hand side. Because the original data are discrete, there are several ties in the similarities and consequently several mergers take place at the same level.

Note that when there are ties, different computer packages may produce different sets of clusters even when using the same clustering method. This is because an arbitrary rule is needed to decide which clusters to merge first. Different rules can result in different sets of clusters being formed (see Morgan and Ray, 1995) .

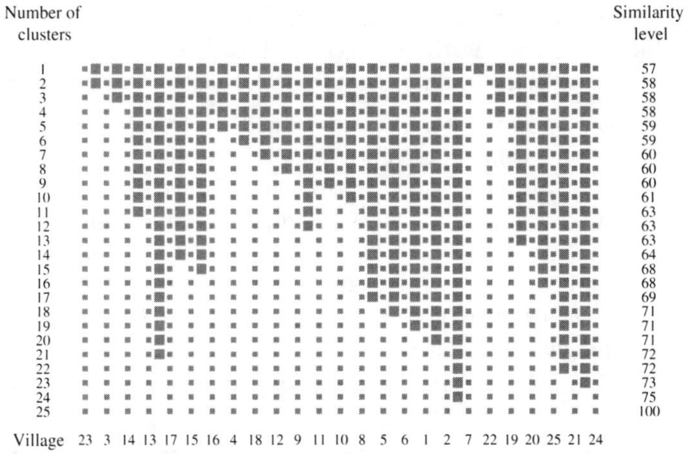

Number of
clusters

Similarity
level

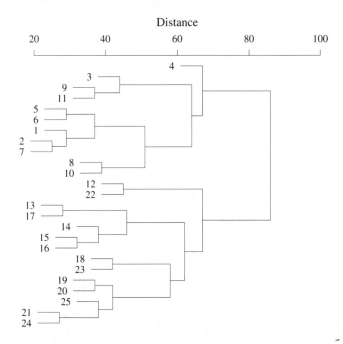

Figure 2.8 *Icicle plot for nearest neighbour clustering, English dialect data*

We may also apply farthest neighbour cluster analysis to these data. This produces the dendrogram in Figure 2.9. Although the dengrograms for the two clustering methods differ in the levels at which clusters merge there is a good measure of agreement between them.

Figure 2.9 *Dendrogram for farthest neighbour (complete linkage) cluster analysis for the English dialect data (distance = 100 − similarity)*

Our interpretations should be read alongside Figure 2.6, which is a map of the East Midlands on which the villages are identified by numbers. Roughly speaking, the identifier codes increase as we move from north to south.

Both methods suggest three fairly well defined clusters. Villages 1, 2, 5, 6, 7, and 8 constitute a first cluster of contiguous villages in Lincolnshire and part of Nottinghamshire. Villages 13, 14, 15, 16, and 17 are all in Leicestershire and Rutland. A group consisting of villages 19, 20, 21, 24, and 25 includes parts of Northamptonshire, Buckinghamshire, and Bedfordshire. There is then a less well defined grouping centred on villages 9 and 11, in south Lincolnshire, but the two methods each append different nearby villages in their neighbourhood. There are several villages which do not fit readily into this scheme or are classified differently by the two methods, in particular, villages 3, 4, 12, 22, and 23.

In order to investigate the position further, we have applied five other methods. These are known as the centroid, median, Baverage, Waverage, and Ward's method, the first and last of which have already been mentioned. The median method is similar to the centroid except that it uses the median instead of the arithmetic mean for the cluster centres. The Baverage method uses the between-cluster average and Waverage uses the within-cluster average. It must be remembered that some of these methods assume that the distances (or similarities in this case) have been constructed from a data matrix of continuous variables which is not the case here. In order to apply the methods, the software needs to convert the similarities to distances and then determine what data matrix would have given rise to them. There is no need to go into the details of this since we are using the methods in a purely empirical fashion to suggest possible clusterings. The results are summarised, along with those for nearest and farthest neighbour methods, in Table 2.11.

The members of the clusters are set out in the columns of the table. Most methods yield four clusters (labelled I, II, III, and V) but with the nearest and farthest neighbour methods we identified small residual clusters which did not seem to fit anywhere else. These are listed as cluster IV. In the case of Ward's method, the geographically northerly and southerly clusters reappear but it fails to distinguish clearly between those in the middle. This middle group has been listed so as to span the columns allocated to clusters II and III. In order to emphasise the common elements in the clusters yielded by the various methods, those that are common are placed on the first line of each cell with the other members on a second line. The cluster membership of villages in brackets is less clear but they have been allocated somewhat arbitrarily to what seemed to be the nearest cluster. Broadly speaking, these further analyses confirm the conclusions we drew from farthest and nearest neighbour methods, but the variations are interesting and should be studied carefully. One particularly interesting case is the only village in Cambridgeshire, village number 23. This does not fit easily into the general scheme of things. Sometimes it is linked with the villages to the west in Northamptonshire and Bedfordshire and sometimes it goes with the more northerly group in Leices-

Table 2.11 *Comparison of the clusters arrived at when various methods are applied to the English dialect data (the numbers refer to the villages)*

| Method | Cluster | | | | |
	I	II	III	IV	V
Nearest	1,2,5,6,7	9,11	13,14,15,16,17		19,20,21,24,25
	8	10,18,12		3,4,(23)	22
Farthest	1,2,5,6,7	9,11	13,14,15,16,17		19,20,21,24,25
	8,(10)	(4)		12,22	18,23
Centroid	1,2,5,6,7	9,11	13,14,15,16,17		19,20,21,24,25
	4,8,10	18,(23)			22
Median	1,2,5,6,7	9,11	13,14,15,16,17		19,20,21,24,25
	3,4	8,10,12	(18,23)		22
Baverage	1,2,5,6,7	9,11	13,14,15,16,17		19,20,21,24,25
	8,(10)	3,(4)	(18,23)		(12)(22)
Waverage	1,2,5,6,7	9,11	13,14,15,16,17		19,20,21,24,25
	4,8,10	3,12	18		22,23
Ward's	1,2,5,6,7	9,11,13,14,15,16,17			19,20,21,24,25
	(4),5,8,10	3,12,18			22,23

tershire. This raises interesting questions about the affinities between various dialects which might be worth pursuing.

In spite of the ambiguities at the edges of the clusters, it is clear that cluster analysis confirms that there are regional groupings in the use of dialect words.

2.6 Comparisons

Faced with such a variety of methods, it is natural to ask whether there are any guiding principles that may be adopted in choosing a method. The problem is that the effectiveness of any method will depend on the "shape" and "location" of any underlying clusters and on the presence or absence of intervening points. These are things we are not likely to know. However, some insight can be gained from looking at the position in two dimensions. Figure 2.10 is designed to illustrate circumstances where the nearest and farthest neighbour methods will differ.

The top panel shows eight points in a plane to be combined into two clusters. The middle panel gives the nearest neighbour clustering. The numbers on the lines joining the points indicate the order in which the points were linked into the cluster, using single linkage. The two clusters consist of a chain snaking from left to right and a single point forming its own cluster at the lower right. The bottom panel gives the farthest neighbour clustering. Again the numbers show the order in which points or clusters were linked together, this time using

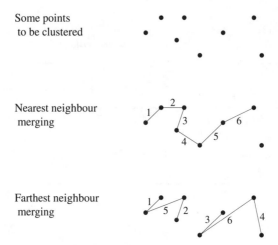

Figure 2.10 *An example showing how points (top panel) are clustered differently using nearest neighbour (middle panel) and farthest neighbour (bottom panel) methods*

complete linkage. Farthest neighbour gives two clusters of equal size, one to the left and one to the right.

Generally, nearest and farthest neighbour give similar results when the clusters are compact and well separated, as in Figure 2.11, but will give different results when the objects are spread out or when there are intermediate objects strung out between the (farthest neighbour) clusters. In Figure 2.11, there are

Figure 2.11 *An arrangement of points that give the same result using either nearest neighbour or farthest neighbour clustering*

three compact clusters with no intermediate points. So, at the next stage of clustering, the same two clusters would merge using either method.

There has been some theoretical work on the question of choice of method. One can start by laying down criteria that any "reasonable" clustering should meet. This approach was taken by Jardine and Sibson (1971) who arrived at the conclusion that the nearest neighbour method was the only one that satisfied all their criteria. However, the middle panel of Figure 2.10 illustrates that the nearest neighbour method may give results that are not very useful. Points close to the existing cluster are successively added to form a set in which the more distant members are so far apart as to raise the question of whether they can be properly regarded as belonging to the same cluster. This phenomenon is known as "chaining", and it can easily lead to non-interpretable clusters.

Although in this chapter we have concentrated on nearest and farthest neighbour cluster analysis, it is advisable in practice to use several methods and compare the results. Where different methods give similar clusters, as in Table 2.11, the analyst can feel confident that they are reflecting some aspect of the data structure. Where they give very different clusters, the analyst may wish to investigate why.

2.7 Clustering variables

There are two types of clustering which can be carried out on a data matrix. One, which has been the subject of this chapter up to this point, is to cluster objects, or individuals. Another type of clustering is clustering variables where we consider whether variables can be grouped because they are distributed similarly across individuals. This duality arises with all analyses that start from a data matrix. If we wished to carry out a cluster analysis on variables we would need measures of similarity between columns of the data matrix instead of between the rows.

In this case, similarity between variables (columns) is typically measured by their correlation. For continuous variables, the product moment correlation will serve very well; for binary or polytomous variables, the tetrachoric and polychoric coefficients respectively, achieve the same end (for more details see Chapter 8).

Alternative measures, such as the Euclidean distance between columns that have not been standardized, might sometimes seem more appropriate for a given problem. This is analogous to carrying out a principal component analysis of the covariance matrix, using unstandardized variables.

2.8 Further examples and suggestions for further work

In this section, we present briefly three further applications of cluster analysis. The first has 21 attributes observed for 24 objects — ancient carvings of archers at Persepolis. The second has responses of 379 people on four binary variables (from the 1986 British Social Attitudes Survey), and the third is an example of clustering educational variables using their correlations. Our analyses are not intended to be complete, and you should explore other methods.

Persian archers

There are twenty-four archers carved in bas-relief going up the south stairs on the west side of the east face of the Apadana at Persepolis in Southern Iran. All of the archers look similar, but they differ in minor details, such as the way the beard curls and the way the head-dress is decorated. Figure 2.12 is a picture of the ninth archer (from the top of the stairs) and identifies 21 features or attributes that may differ between archers. Each attribute has only a small number of variants, usually two or three. Details are given in Roaf (1983).

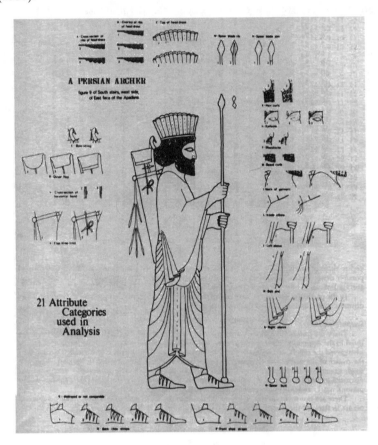

Figure 2.12 *Archer number 9 at Persepolis, surrounded by variants of 21 attributes (reproduced from Roaf 1978 with permission)*

Table 2.12 gives the data matrix showing which archer has which variant of each attribute. The attributes are labelled A to U and variants of an attribute are labelled by integers between 1 and 6. A zero in the table indicates that that attribute is missing from the bas-relief of that archer, due to damage over time.

Archaeologists and art historians are interested in whether the archers were carved by a single sculptor or by several, and in whether these data suggest any groupings that might bear on this question. Roaf (1978,1983) reported several cluster analyses identifying clusters of archers which might have been carved by the same sculptor or team of sculptors. We use these data to illustrate farthest neighbour cluster analysis, and we invite you to try other methods.

Table 2.12 *Data matrix showing for each of 24 archers which variant of each of 21 attributes A to U he possesses (a zero indicates that the attribute for that archer is missing)*

Archer	Attribute																				
	A	B	C	D	E	F	G	H	I	J	K	L	M	N	O	P	Q	R	S	T	U
1	2	2	1	2	3	2	1	1	2	1	1	3	0	0	2	3	4	2	2	2	0
2	2	2	1	2	2	2	1	1	2	1	1	3	1	1	2	0	4	2	2	1	2
3	2	2	1	2	2	2	1	1	2	1	2	3	2	2	2	3	0	3	2	1	3
4	2	2	1	2	3	1	1	1	2	1	2	3	2	2	2	3	4	2	2	1	3
5	2	3	1	3	2	2	1	1	2	1	2	0	2	2	4	3	4	3	2	1	3
6	2	3	1	2	2	2	1	1	2	1	2	3	2	2	4	3	4	3	2	1	3
7	2	3	1	3	3	2	1	1	2	2	2	3	2	2	3	3	4	3	2	1	3
8	2	3	1	3	3	2	2	1	2	2	1	2	2	2	3	3	4	3	2	1	3
9	3	1	1	3	2	2	1	2	2	1	1	2	1	1	3	2	2	4	3	2	2
10	3	1	1	3	2	2	1	2	2	2	1	2	2	2	3	2	2	2	2	2	2
11	3	1	1	3	2	2	1	2	2	1	2	2	2	2	0	2	2	4	3	2	2
12	3	1	2	2	2	2	1	2	2	1	2	2	1	1	1	5	2	2	3	1	3
13	3	1	1	3	2	2	1	2	2	2	2	2	1	1	0	0	3	2	2	2	3
14	3	1	1	3	2	2	1	2	2	2	2	2	2	2	5	3	2	2	2	2	3
15	3	1	1	3	2	2	1	2	2	1	2	2	2	2	3	2	2	4	3	2	2
16	3	1	1	3	2	2	1	2	1	1	2	2	1	1	3	2	4	4	3	2	2
17	3	1	2	3	2	3	1	2	2	2	2	2	1	2	2	3	4	2	2	2	2
18	3	1	1	3	2	2	1	2	2	1	2	2	2	2	6	1	1	2	2	2	2
19	2	1	1	3	2	2	1	2	2	1	2	2	2	2	3	1	4	2	2	2	2
20	2	1	2	2	2	2	3	1	2	1	2	2	2	2	1	3	4	2	2	1	3
21	2	1	2	2	2	2	3	1	2	1	2	2	2	2	0	3	4	2	2	2	3
22	2	1	2	2	2	2	3	1	1	1	2	2	2	1	0	3	4	2	2	2	3
23	2	1	2	2	2	2	3	1	2	1	2	2	2	1	1	3	4	2	2	2	3
24	2	1	1	2	2	2	3	1	2	1	2	2	2	1	2	3	4	2	2	2	3

Table 2.13 gives a table of similarities for all pairs of archers. The similarity of a pair of archers is calculated from Table 2.12 to be the proportion of attributes for which each archer has the same variant, multiplied by 21. For a pair of archers with no attributes missing, this is just the count of the number of attributes for which they share the same variant. For two archers with m missing attributes from one or both of them, the similarity is the corresponding count but scaled up by a factor of $21/(21 - m)$. For example, the similarity between archers 1 and 2 is $15 \times 21/17 = 18.5$ because attributes M, N, P, and U are missing for one or other archer and there are 15 matches among the remaining 17 attributes. The rationale for this scaling up is that archers with several features missing would otherwise have misleadingly low similarities with the other archers.

Table 2.13 *Table of similarities for 24 Persian archers*

Archer	1	2	3	4	5	6	7	8	9	10	11	12	13	14	15	16	17	18	19	20	21	22	23	24
1	21																							
2	18.5	21																						
3	16.1	15.5	21																					
4	17.5	14.7	17.8	21																				
5	12.4	12.2	17.7	14.0	21																			
6	14.0	13.6	18.9	16.0	20.0	21																		
7	12.8	10.5	15.8	15.0	17.8	17.0	21																	
8	11.7	9.4	12.6	12.0	15.8	14.0	18.0	21																
9	8.2	10.5	6.3	4.0	7.4	6.0	6.0	7.0	21															
10	9.3	9.4	8.4	7.0	9.4	8.0	10.0	11.0	16.0	21														
11	7.4	7.7	9.9	7.0	11.1	9.0	8.0	7.0	17.0	16.0	21													
12	7.0	10.5	9.4	8.0	8.4	9.0	6.0	5.0	13.0	10.0	12.6	21												
13	9.2	9.3	9.3	7.0	10.5	8.0	9.0	8.0	13.0	14.0	13.3	13.0	21											
14	9.3	8.4	11.6	10.0	11.6	10.0	11.0	10.0	11.0	16.0	14.7	12.0	18.8	21										
15	7.0	7.4	9.4	7.0	10.5	9.0	9.0	8.0	18.0	17.0	21.0	12.0	13.3	14.0	21									
16	7.0	9.4	6.3	5.0	8.4	7.0	7.0	6.0	18.0	13.0	16.8	12.0	14.4	11.0	17.0	21								
17	9.3	9.4	8.4	9.0	9.4	8.0	9.0	8.0	11.0	14.0	12.6	11.0	15.5	15.0	12.0	12.0	21							
18	9.3	9.4	10.5	9.0	11.6	10.0	9.0	8.0	13.0	16.0	16.8	11.0	15.5	16.0	16.0	13.0	14.0	21						
19	11.7	11.6	11.6	11.0	13.6	12.0	12.0	11.0	13.0	16.0	15.8	10.0	14.4	15.0	16.0	14.0	14.0	18.0	21					
20	11.7	11.6	14.7	14.0	14.7	15.0	12.0	12.0	6.0	9.0	9.4	13.0	9.9	11.0	9.0	7.0	11.0	11.0	13.0	21				
21	13.6	11.1	14.4	13.0	14.4	14.0	11.0	11.0	7.0	10.0	10.5	11.0	11.1	12.0	10.0	8.0	12.0	12.0	14.0	19.0	21			
22	12.4	11.1	12.2	11.0	12.2	12.0	9.0	9.0	7.0	8.0	8.4	11.0	11.1	10.0	8.0	10.0	10.0	10.0	12.0	17.0	18.9	21		
23	12.8	11.6	12.6	12.0	12.6	13.0	10.0	10.0	8.0	9.0	9.4	13.0	12.2	11.0	9.0	9.0	11.0	11.0	13.0	19.0	20.0	20.0	21	
24	15.2	13.6	14.7	14.0	13.6	14.0	11.0	11.0	9.0	10.0	10.5	11.0	13.3	13.0	10.0	10.0	11.0	12.0	14.0	17.0	18.9	18.9	19.0	21

Figure 2.13 gives the dendrogram using farthest neighbour clustering. This shows two major clusters. The first consists of archers 9 to 19 in the middle section of the staircase and the second consists of archers 1 to 8 at the top and archers 20 to 24 at the bottom. The major clusters are divided into smaller

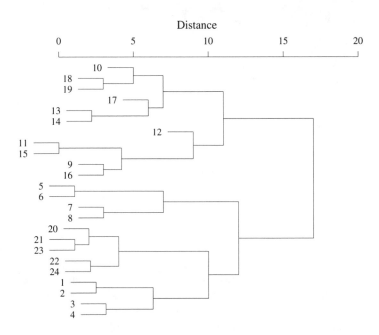

Figure 2.13 *Dendrogram for farthest neighbour (complete linkage) cluster analysis for the Persian archers data (distance = 21 − similarity)*

groups of (mainly) contiguous archers — in particular the bottom five archers (20 to 24) form a tight cluster. Within this overall pattern, there are two relatively isolated archers, 12 and 17. By examining the 12th and 17th rows and columns of the similarity matrix (Table 2.13), you can confirm that all of their similarities are small.

You could try other methods of cluster analysis on these data and compare the results. You might also look at Section 3.7 where the same data are analyzed using multidimensional scaling.

An alternative visual presentation of the archers and their similarities is shown in Figure 2.14. This diagram, reproduced from Roaf (1978) (Figure 5), shows the positions on the staircase and links individuals with 15 or more attribute variants in common. Here three groups of archers are clearly identified, the bottom five being very similar. This simple analysis reinforces the main conclusions from the dendrogram and shows that a customised analysis and presentation can sometimes provide a better summary of the similarity structure than standard cluster analysis.

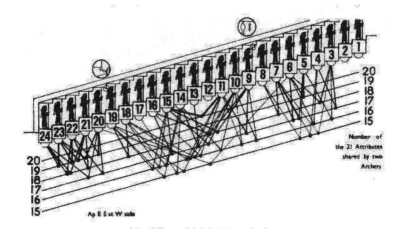

Figure 2.14 *A diagram of archers on the staircase linking those with 15 or more attribute variants in common (reproduced from Roaf 1978 with permission)*

Attitude to abortion

This is an instructive example of a less successful application of cluster analysis. Section 7.1 describes some data that have been extracted from the 1986 British Social Attitudes Survey. Binary responses to four items are given for 379 respondents. The items are questions relating to the circumstances under which a respondent would consider that an abortion should be allowed under law. The four circumstances are (briefly): the woman decides on her own that she does not wish to have the child; the couple agree that they do not want to have the child; the woman is not married and does not wish to marry the man; and the couple cannot afford to have any more children. Table 2.14 shows the numbers of people responding positively (the law should al-

Table 2.14 *Attitude to abortion marginal frequencies for four binary variables*

		Response		
Variable	Item	1	0	Total
x_1	WomanDecide	166	213	379
x_2	CoupleDecide	225	154	379
x_3	NotMarried	241	138	379
x_4	CannotAfford	234	145	379

low an abortion) and negatively (an abortion should not be allowed) for each item, together with a short item name. Positive responses are coded as 1 and negative responses as 0.

Table 2.15 sets out the 16 possible response patterns and their frequencies. The patterns are arranged by the number of 1s they contain. The data may be regarded as 379 observations on four binary variables x_1, x_2, x_3, and x_4. Thus,

Table 2.15 *Response patterns, frequencies and cluster allocation using farthest neighbour (complete linkage) clustering for the attitude to abortion data*

Pattern	x_1 x_2 x_3 x_4	Total	Frequency	Cluster
1	0 0 0 0	0	103	1
2	0 0 0 1	1	13	1
3	0 0 1 0	1	10	2
4	0 1 0 0	1	9	1
5	1 0 0 0	1	1	1
6	0 0 1 1	2	21	2
7	0 1 0 1	2	6	1
8	1 0 0 1	2	0	1
9	0 1 1 0	2	7	2
10	1 0 1 0	2	0	2
11	1 1 0 0	2	3	1
12	0 1 1 1	3	44	2
13	1 0 1 1	3	6	2
14	1 1 0 1	3	3	1
15	1 1 1 0	3	12	2
16	1 1 1 1	4	141	2
		Total	379	

for example, 103 respondents thought an abortion should not be allowed in any of the four situations; 13 thought an abortion should be allowed if the couple could not afford have the child; and so on to the last row, where 141 respondents thought that an abortion should be allowed in all four situations.

Usually in a cluster analysis, we begin with n objects or individuals. But here the 379 people are already grouped according to which of the 16 response patterns they have. So now we will try to cluster the response patterns (and thereby implicitly to cluster individuals). For example, we may find evidence that people fall into two groups (essentially pro- or anti-abortion), or that attitudes are more diverse.

As there are only four binary variables, the similarity measure $r = (a+d)/p$ has just four distinct values $(0, \frac{1}{4}, \frac{1}{2}$ and $\frac{3}{4})$ corresponding to whether two patterns match on 0, 1, 2 or 3 responses. This could only give very limited possibilities for clustering response patterns. Instead, we use the similarities shown in Table 2.16. and which we explain below.

Table 2.16 *Similarities between response patterns for the attitude to abortion data*

Pattern	1	2	3	4	5	6	7	8	9	10	11	12	13	14	15	16
0000	9.60	6.99	6.85	7.14	7.82	4.24	4.53	5.21	4.39	5.07	5.36	1.78	2.46	2.75	2.61	0
0001	6.99	8.61	4.24	4.53	5.21	5.86	6.15	6.83	1.78	2.46	2.75	3.40	4.08	4.37	0	1.62
0010	6.85	4.24	8.43	4.39	5.07	5.81	1.78	2.46	5.97	6.65	2.61	3.35	4.03	0	4.19	1.57
0100	7.14	4.53	4.39	8.82	5.36	1.78	6.21	2.75	6.08	2.61	7.04	3.46	0	4.43	4.30	1.68
1000	7.82	5.21	5.07	5.36	10.10	2.46	2.75	7.49	2.61	7.36	7.64	0	4.74	5.03	4.90	2.28
0011	4.24	5.86	5.81	1.78	2.46	7.43	3.40	4.08	3.35	4.03	0	4.97	5.65	1.62	1.57	3.19
0101	4.53	6.15	1.78	6.21	2.75	3.40	7.83	4.37	3.46	0	4.43	5.08	1.62	6.05	1.68	3.30
1001	5.21	6.83	2.46	2.75	7.49	4.08	4.37	9.11	0	4.74	5.03	1.62	6.36	6.65	2.28	3.90
0110	4.39	1.78	5.97	6.08	2.61	3.35	3.46	0	7.65	4.19	4.30	5.04	1.57	1.68	5.87	3.26
1010	5.07	2.46	6.65	2.61	7.36	4.03	0	4.74	4.19	8.93	4.90	1.57	6.32	2.28	6.47	3.86
1100	5.36	2.75	2.61	7.04	7.64	0	4.43	5.03	4.30	4.90	9.33	1.68	2.28	6.71	6.58	3.97
0111	1.78	3.40	3.35	3.46	0	4.97	5.08	1.62	5.04	1.57	1.68	6.66	3.19	3.30	3.26	4.88
1011	2.46	4.08	4.03	0	4.74	5.65	1.62	6.36	1.57	6.32	2.28	3.19	7.94	3.90	3.86	5.48
1101	2.75	4.37	0	4.43	5.03	1.62	6.05	6.65	1.68	2.28	6.71	3.30	3.90	8.33	3.97	5.59
1110	2.61	0	4.19	4.30	4.90	1.57	1.68	2.28	5.87	6.47	6.58	3.26	3.86	3.97	8.15	5.54
1111	0	1.62	1.57	1.68	2.28	3.19	3.30	3.90	3.26	3.86	3.97	4.88	5.48	5.59	5.54	7.16

The similarities in Table 2.16 are obtained by weighting each response for each variable to give the following measure of similarity between patterns i and j:

$$\sum_{k=1}^{4} w_{k1} x_{ik} x_{jk} + \sum_{k=1}^{4} w_{k0}(1 - x_{ik})(1 - x_{jk}),$$

where x_{ik} and x_{jk} are the responses to item k for patterns i and j, respectively, and where w_{k1} and w_{k0} are weights. We use $w_{k1} = n/n_k$ and $w_{k0} = n/(n-n_k)$ where n_k is the number of positive responses for item k and $n - n_k$ is the number of negative responses. The product $x_{ik} x_{jk}$ will equal 1 if x_k equals 1 in both patterns, and will be zero otherwise. Likewise $(1 - x_{ik})(1 - x_{jk})$ will equal 1 if x_k equals 0 in both patterns. So using weights $w_{k1} = n/n_k$ and $w_{k0} = n/(n - n_k)$ has the effect of giving more weight to agreement on rarer responses.

For example, from Table 2.15, response patterns 6 and 7 both agree on $x_1 = 0$ and $x_4 = 1$. So $(1 - x_{61})(1 - x_{71}) = 1$ and $x_{64} x_{74} = 1$ are the only non-zero terms in the above formula. From Table 2.14, the relative frequency of 0 for item 1 is $213/379$ and the relative frequency of 1 for item 4 is $234/379$. Hence, the similarity between patterns 6 and 7 is

$$\frac{379}{213} \times 1 \; + \; \frac{379}{234} \times 1 \; = \; 3.40 \,.$$

Likewise for patterns 7 and 8, which agree on $x_3 = 0$ and $x_4 = 1$, the similarity is

$$\frac{379}{138} \times 1 \; + \; \frac{379}{234} \times 1 \; = \; 4.37 \,.$$

Thus, patterns 7 and 8 are judged to be more similar than patterns 7 and 6, because $x_3 = 0$ is a rarer response than $x_1 = 0$.

You could consider the appropriateness of this measure of similarity which gives a larger weight to less frequent responses than to more frequent responses. Another feature of this measure is that the diagonal elements of the similarity matrix are not all equal, but, of course, they do not affect the clustering process.

Figure 2.15 shows the dendrogram for nearest neighbour (single linkage) cluster analysis using the similarity matrix in Table 2.16. This is an extreme example of *chaining*. The first two patterns that merge, numbers 1 and 5, have pattern 11 as their nearest neighbour so 11 then merges. This process continues, with each pattern successively merging into the same group. Only a single link is required to add a new object (response pattern) to the existing cluster, resulting in a single chain.

Figure 2.16 shows the dendrogram for farthest neighbour (complete linkage) clustering. This looks more successful. Here a new pattern can only be joined to a cluster if it is close to *all* members of that cluster, that is, complete linkage to every member is required.

In Figure 2.16, two clusters each of eight patterns can be identified. Cluster 1 (in the top half of the dendrogram) consists of response pattern 1 (0000) and other patterns with several 0s, and cluster 2 (in the bottom half) has pat-

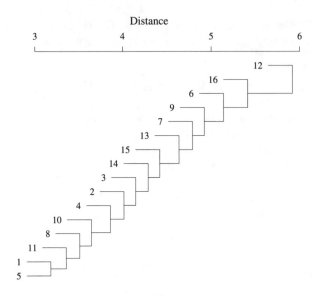

Figure 2.15 *Dendrogram for nearest neighbour (single linkage) cluster analysis for the attitude to abortion data (distance = 11 − similarity)*

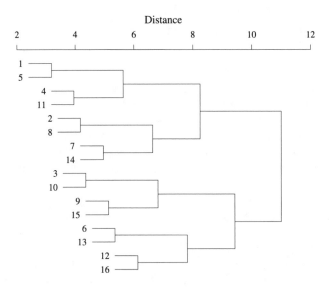

Figure 2.16 *Dendrogram for farthest neighbour (complete linkage) cluster analysis for the attitude to abortion data (distance = 11 − similarity)*

tern 16 (1111) and other patterns with several 1s. This allocation of patterns to clusters is listed in the last column of Table 2.15. So perhaps respondents in cluster 1 have a generally unfavourable attitude to abortion while those in cluster 2 have a more favourable attitude. But there are two exceptions.

Pattern 3 (0010) has been allocated to cluster 1 and pattern 14 (1101) to cluster 2. Closer inspection reveals that all patterns in cluster 1 have $x_3 = 0$ and all in cluster 2 have $x_3 = 1$. The division into these two clusters could therefore have been achieved on the basis of a single item relating to whether the woman is married. Does this make sense? To what extent is it the result of our choice of similarity measure?

Table 9.3 shows the result of latent class analysis on these same data. The model-based method of cluster analysis used there seems to have been more successful. In particular, all patterns with three or more 0s are allocated to class 1 and all with three or more 1s are allocated to class 2. Some patterns with two 1s and two 0s are allocated to class 1 and others to class 2.

Clustering educational variables

Rather than clustering individuals or objects, this last example aims to examine how five measurements made on secondary school girls in 1964 relate to four measurements (three the same and one new) made on the same girls in 1968 (see Peaker 1971). The data set, which we will refer to as "educational circumstances", comes from a national survey of primary school children in 1964 and a follow-up survey in 1968. About one quarter of the children could not be traced which could introduce bias as the missing children may differ from those who were followed up. We use data for 398 girls in their final year of primary school in 1964 and in their fourth year of secondary school in 1968. The nine variables we analyze are composite measures described in Table 4 of Peaker (1971). The variables for 1964 are (briefly): parental circumstances (x_1), details of class teacher (x_2), school-parent interaction (x_3), girl's attitude (x_4) and test score (x_5), and in 1968: type of school (x_6), parental circumstances (x_7), school-parent interaction (x_8), and test score (x_9). The correlations given in Table 2.17 are taken from Table 7 of Peaker (1971) with one correction to an obvious typographical error.

Table 2.17 *Correlation matrix, educational circumstances*

	x_1	x_2	x_3	x_4	x_5	x_6	x_7	x_8	x_9
x_1	1								
x_2	0.177	1							
x_3	0.305	0.155	1						
x_4	0.193	0.124	0.243	1					
x_5	0.501	0.134	0.556	0.317	1				
x_6	0.423	0.124	0.308	0.308	0.572	1			
x_7	0.770	0.184	0.351	0.193	0.436	0.388	1		
x_8	0.206	−0.050	0.149	0.128	0.252	0.382	0.206	1	
x_9	0.499	0.127	0.413	0.339	0.758	0.613	0.459	0.315	1

It is interesting to see whether a cluster analysis on the correlation matrix will combine measurements made at the same time, or measurements of the same characteristic made at different times, or a mixture of these. A further question is which other variables might be associated with success in the two tests. Before proceeding with this analysis, we make two cautionary remarks. First, a cluster analysis is not an ideal way of investigating how one set of variables depends on another, and secondly, the use of product moment correlation as a measure of association between variables some of which are ordinal could be improved upon (see Section 8.5). However, the use of simple descriptive methods may sometimes reveal interesting aspects that a more focused analysis might miss.

The dendrogram in Figure 2.17 shows the result of a nearest neighbour cluster analysis and gives the short names of the variables. Two pairs of variables,

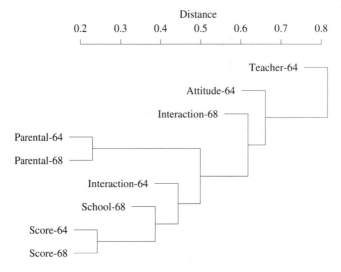

Figure 2.17 *Dendrogram for nearest neighbour (single linkage) cluster analysis for the correlation matrix in Table 2.17 (distance = 1 − correlation)*

parental circumstances in 1964 and in 1968, and total test scores 1964 and 1968, are each closely linked but the two measurements for school-parent interaction are linked only at the sixth out of eight steps. There is some chaining of variables, but to determine whether this is due to the use of single linkage or whether it would still appear with other methods we leave as an exercise for you. Overall, from the nearest neighbour analysis, we might conclude that the teacher's characteristics, the girl's attitude in 1964, and school-parent interaction in 1968 are only weakly associated with the test scores, whereas the other four variables have stronger associations with the test scores. You can confirm these conclusions by examining the correlation matrix.

Finally

The analyses reported in this chapter have mainly used either nearest or farthest neighbour cluster analysis. You are recommended to try out other methods on these data sets and on some of the other data sets given in later chapters. You should also compare the results of the analyses in this chapter with analyses using some of the same data sets in Section 3.7 (archers and English dialect data), Section 5.9 (educational circumstances), Chapter 7 (attitude to abortion data), and especially the latent class analysis in Section 9.2 (attitude to abortion data).

2.9 Further reading

Everitt, B. S., Landau, S., and Leese, M. (2001). *Cluster Analysis.* 4th edition. London: Arnold.

Everitt, B. S. and Rabe-Hesketh, S. (1997). *The Analysis of Proximity Data.* London: Arnold.

Gordon, A. D. (1999) *Classification.* 2nd edition. London: Chapman and Hall/CRC.

CHAPTER 3

Multidimensional Scaling

3.1 Introduction

Multidimensional scaling is one of several multivariate techniques that aim to reveal the structure of a data set by plotting points in one or two dimensions. The basic idea can be motivated by a geographical example. Suppose we are given the distances between pairs of cities and are asked to reconstruct the two-dimensional map from which those distances were derived. We could attempt to do this by a process of trial and error by moving points about on a sheet of paper until we got the distances right. A procedure that does this automatically is called multidimensional scaling (MDS). The "multi" part of the name refers to the fact that we are not restricted to constructing maps in one or two dimensions.

This simple example differs in two important ways from the typical MDS problem. In the first place, there is no ambiguity about what we mean by the "distance" between two cities (measured in miles or kilometers in a straight line), whereas in the typical MDS problem there is often a degree of arbitrariness in the definition of distance which, in some cases, may be based on subjective assessments rather than precise measurement. Secondly, we know that the cities can be located on a two-dimensional map (provided that the curvature of the earth and other topographical features can be ignored), whereas in the typical MDS problem we would have little idea how many dimensions would be necessary in order to reproduce, even approximately, the given distances between objects of interest. Indeed one of the prime objects of the analysis will be to discover whether such a representation is possible in a small number of dimensions. Unless this can be done, preferably in one or two dimensions, we shall not be able to take advantage of the eye's ability to spot patterns in the plots. Even if it turns out that more than two dimensions are necessary, the main way we can view the points is by projecting them onto two-dimensional space.

The input data for MDS is in the form of a distance matrix representing the distances between pairs of objects. We have already discussed the construction of such matrices in Chapter 2 and there is nothing to add here. However, whereas the choice between distance and proximity was largely a matter of indifference in cluster analysis, distance is the prime concept in MDS. Thus although we may start with a proximity or similarity matrix, it may need to be converted to a distance matrix in the course of the analysis; the output will be expressed in terms of distance.

As we have said, MDS is used to determine whether the distance matrix
may be represented by a map or configuration in a small number of dimensions
such that distances on the map reproduce, approximately, the original distance
matrix $\{\delta_{ij}\}$. For example, we would aim to have the two objects that are
closest together according to the distance matrix closest together on the map,
and so on. As we have posed the problem, the distances on the map would
be in the same metric (scale of measurement) as the original δ_{ij}'s. This is
often known as *classical* multidimensional scaling. However, it is often the
case, particularly in social science research, that the values of the δ_{ij}'s may be
interpreted only in an ordinal sense as if, for example, the distances come from
subjective similarity ratings. In such cases, it may be more reasonable only to
attempt to produce a map on which the distances have the right rank order.
This is called *ordinal* or *non-metrical* multidimensional scaling. In this chapter,
we shall be mainly concerned with ordinal MDS. In the second example in
Section 3.2 below, students were asked to rate the degree of similarity between
pairs of countries on a nine-point scale. Similarity, here, is a subjective thing
for which there is no natural underlying "space" reflected in the similarities.
Part of the interest in the analysis is to try to uncover which attributes of the
countries appear to carry weight in the students' judgement of similarity.

Returning to classical scaling, suppose that we have four cities labelled A,
B, C, and D and that the distances (in hundreds of miles) between the pairs
of cities are as given by the following matrix:

$$
\begin{array}{c}
A \\ B \\ C \\ D
\end{array}
\left(
\begin{array}{cccc}
- & & & \\
2 & - & & \\
1 & 3 & - & \\
5 & 3 & 6 & -
\end{array}
\right)
$$

Using multidimensional scaling (or by inspection), it is possible to represent
this distance matrix exactly in one dimension. A possible solution is given in
Figure 3.1.

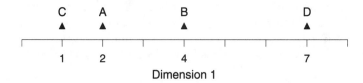

Figure 3.1 *A one-dimensional configuration of four cities using classical MDS*

We shall denote the distance between objects i and j in the above configuration by d_{ij} and in this case, these distances are precisely equal to the δ_{ij}'s.
In classical MDS, we seek a configuration such that the d_{ij}'s, the inter-point
distances in the configuration, will be approximately equal to the corresponding δ_{ij}'s, as given in the distance matrix; whereas in ordinal MDS, the object
is only to find a configuration such that the d_{ij}'s are in the same rank order
as the corresponding δ_{ij}'s.

Given the Euclidean distances between n objects, it is always mathematically possible to find a configuration in $(n-1)$ dimensions that matches perfectly, but this would be of little use. Our aim will be to obtain a fairly good *approximate* representation in a small number of dimensions.

Measures of similarity between variables

We have already remarked in Section 2.7 that one can reverse the roles of variables and objects. Instead of clustering objects, which was our main concern, we could have clustered variables. This duality arises with all analyses that start from a data matrix. If we wished to carry out an MDS analysis on variables, we would need measures of similarity between columns of the data matrix instead of between the rows.

3.2 Examples

Reproducing a two-dimensional map from air distances between pairs of cities

MDS was carried out to determine whether a two-dimensional map could be produced from a matrix of pairwise distances between ten cities in Europe and Asia. The dissimilarity or distance matrix is shown in Table 3.1.

Table 3.1 *Distances between ten cities in air miles*

	London	Berlin	Oslo	Moscow	Paris	Rome	Beijing	Istanbul	Gibraltar	Reykjavik
London	–									
Berlin	570	–								
Oslo	710	520	–							
Moscow	1550	1000	1020	–						
Paris	210	540	830	1540	–					
Rome	890	730	1240	1470	680	–				
Beijing	5050	4570	4360	3600	5100	5050	–			
Istanbul	1550	1080	1520	1090	1040	850	4380	–		
Gibraltar	1090	1450	1790	2410	960	1030	6010	1870	–	
Reykjavik	1170	1480	1080	2060	1380	2040	4900	2560	2050	–

The solution from a classical MDS in two dimensions is shown in Figure 3.2.

The MDS has mapped points in two-dimensional space such that the "straight line" (Euclidean) distances between the points d_{ij} match the observed distances δ_{ij}. The d_{ij}'s are very close to the (rescaled) δ_{ij}'s. They are not precisely equal because the δ_{ij}'s are not "straight line" distances but distances across the surface of a sphere.

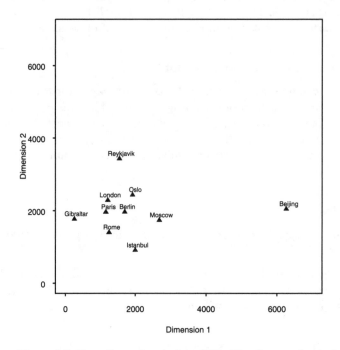

Figure 3.2 *Two-dimensional plot of 10 cities from a classical MDS*

Figure 3.2 is recognizable as a map of Europe and Asia. However, in general a configuration may need to be rotated and/or reflected in order to clarify the interpretation. Three important points about interpreting MDS solutions are:

i) The configuration can be reflected without changing the inter-point distances

ii) The inter-point distances are not affected if we change the origin by adding or subtracting a constant from the row or the column coordinates

iii) The set of points can be rotated without affecting the inter-point distances. This comes to the same thing as rotating the axes

We must therefore be prepared to look for the most meaningful set of axes when interpreting an MDS solution. This idea will become clearer when we come to the next example. To summarise, the interpretation we put upon any MDS solution must be invariant under reflection, translation, and rotation.

An attempt to determine the dimensions underlying similarity judgements for pairs of 12 countries

In 1968, a group of 18 students was asked to rate the degree of similarity between each pair of 12 countries on a scale from 1 ("very different") to 9 ("very similar"). The study is described in Kruskal and Wish (1994), but our analysis is slightly different. The mean similarity ratings were calculated across students to obtain the similarity matrix in Table 3.2.

Table 3.2 *Subjective similarities between pairs of 12 countries*

	Brazil	Congo	Cuba	Egypt	France	India	Israel	Japan	China	Russia	USA	Yugo-slavia
Brazil	–											
Congo	4.83	–										
Cuba	5.28	4.56	–									
Egypt	3.44	5.00	5.17	–								
France	4.72	4.00	4.11	4.78	–							
India	4.50	4.83	4.00	5.83	3.44	–						
Israel	3.83	3.33	3.61	4.67	4.00	4.11	–					
Japan	3.50	3.39	2.94	3.83	4.22	4.50	4.83	–				
China	2.39	4.00	5.50	4.39	3.67	4.11	3.00	4.17	–			
Russia	3.06	3.39	5.44	4.39	5.06	4.50	4.17	4.61	5.72	–		
USA	5.39	2.39	3.17	3.33	5.94	4.28	5.94	6.06	2.56	5.00	–	
Yugo-slavia	3.17	3.50	5.11	4.28	4.72	4.00	4.44	4.28	5.06	6.67	3.56	–

Ordinal MDS was applied to this similarity matrix, because the similarities are based on subjective judgements. The resulting solution in two dimensions is shown in Figure 3.3 below.

We have to consider whether we can identify what is varying as we move along the two axes. Thus, for example, what do those countries on the right of the diagram have more of than those on the left, or those at the top than those at the bottom? Nothing very obvious seems to emerge from such comparisons but we must remember that the orientation is arbitrary and maybe the message will be clearer if we consider other rotations. The dotted axes shown on Figure 3.3 correspond to a rotation that does seem to have an interpretation in terms of meaningful variables. Kruskal and Wish (1994), p. 326, note that variation in the direction of the axis that runs from bottom left to top right corresponds to a tendency to be pro-Western or pro-Communist. Those at the top right are the more pro-Communist and those at the bottom left are the more pro-Western. Variation in the direction at right angles separates the developed (top left) from the developing (bottom right) countries. It thus appears that when making their judgements, the students were taking account, consciously or unconsciously, of two types of difference, and the analysis has helped us to identify what those two dimensions were.

It is worth adding two cautionary remarks about this example. The similarities were obtained by averaging the assessments of the 18 students. Implicitly, therefore, we are assuming that all are using the same two dimensions and that they are giving them the same relative weight. This may not be the case and it would be useful to have a method of discovering whether this was

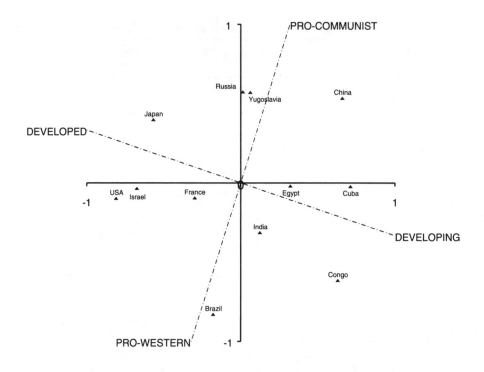

Figure 3.3 *Two-dimensional plot of countries from ordinal MDS*

true. Such methods, known as Individual Scaling or Three-Way Scaling, are available but are outside the scope of this book (see, for example, Borg and Groenen (1997) or Kruskal and Wish (1994)) .

The second remark is that the identification of interpretable axes for a plot is not always the best way of discerning interesting patterns. It may be that we can identify clusters of points which have practical significance, as in the acoustic confusion example in Section 3.7, or, as in the colour data example in Section 3.6, the clue may be in the "horseshoe" shape of the two-dimensional plot.

3.3 Classical, ordinal, and metrical multidimensional scaling

We now pose the problem of multidimensional scaling in more formal terms so that we can outline the algorithms used to arrive at a solution.

Classical scaling

In *classical* MDS, the aim is to find a configuration in a low number of dimensions such that the distances between the points in the configuration, d_{ij}, are close in value to the observed distances δ_{ij}. The method treats the distances as Euclidean distances. We saw in Chapter 2 how to go from a data matrix to a Euclidean distance matrix; here we have to go in the reverse direction and recover the data matrix from the distances. We cannot recover everything because information about location and orientation is lost in the process of calculating distances, but we can determine the configuration. This problem can be tackled algebraically, and it turns out that the solution gives us a series of approximations starting with one dimension, then two, and so on. It also happens, however, that the mathematics involved is equivalent to that for another problem for which the solution is already known. This establishes an interesting link with principal components analysis that we shall discuss in Chapter 5. We shall return to this link in that chapter but we can prepare the ground by expressing the classical MDS problem in a slightly different way. If we start with an $n \times p$ data matrix, we first construct a distance table and then might seek to find a two- or three-dimensional map on which the inter-point distances are as close as possible to the original distances. Another way of putting this is to say that we are looking for a new data matrix, with two or three columns, which is close to the original matrix in the sense that it gives rise to (nearly) the same distance matrix.

Having found a solution, we may wish to have a measure of how good the fit is. This would be particularly useful for helping us to judge how many dimensions are necessary to get a good enough fit. An obvious way to do this is to look at the sum of squares $\sum_{i<j}(d_{ij} - \delta_{ij})^2$. (This is mathematically appropriate since the the fits obtained are best in a least squares sense.) However, the simple sum of squares depends on the scale in which the distances are measured. It is, therefore, preferable to normalize the sum of squares and, in order to reduce it to the same units as the distances, to take the square root. Our goodness-of-fit measure is then

$$\sqrt{\frac{\sum_{i<j}(d_{ij} - \delta_{ij})^2}{\sum_{i<j} d_{ij}^2}}. \tag{3.1}$$

This measure is called the *stress* or, sometimes, the normalised stress. There are other ways of calculating a normalised stress measure. For example, an alternative measure of stress may be obtained by replacing d_{ij} with δ_{ij} in the denominator of (3.1). Values of stress that are close to zero would indicate that the MDS solution is a good fit to the original δ_{ij}'s.

Ordinal (non-metrical) scaling

Very often it is not the actual value of δ_{ij} that is important or meaningful, but
its value in relation to the distances between other pairs of objects. This is
particularly true when the δ_{ij}'s are the result of an experiment where subjects
are asked to give their subjective assessments of the distance between objects.
In such cases, the δ_{ij}'s can be interpreted only in an ordinal sense. In *ordinal*
MDS, the aim is to find a configuration such that the d_{ij}'s are in the same
rank order as the original δ_{ij}'s. So, for example, if the distance apart of objects
1 and 3 rank fifth among the δ_{ij}'s then they should also rank fifth in the MDS
configuration. The emphasis in this chapter, as noted in Section 3.1, is on
ordinal MDS.

In ordinal MDS, we construct fitted distances, often called *disparities*, $\hat{d}_{ij}$,
from the d_{ij}'s such that the $\hat{d}_{ij}$'s are in the same rank order as the δ_{ij}'s. We
can think of the $\hat{d}_{ij}$'s as "smoothed" versions of the d_{ij}'s. This smoothing
process is carried out using a method called least-squares monotonic regression ("monotonic" means that the regression curve is either non-decreasing
or non-increasing). Using this method, the d_{ij}'s are regressed on the δ_{ij}'s. In
a plot of d_{ij} versus δ_{ij}, we would like to see a monotonic curve (one where
the lines joining adjacent points are flat/decreasing if δ_{ij} are similarities or
flat/increasing if δ_{ij} are dissimilarities). If the d_{ij}'s and the δ_{ij}'s have the
same rank order, then the plot will show such a monotonic curve and the
d_{ij}'s will not require any smoothing. Usually, however, there will be some departures from monotonicity and some smoothing will be necessary. The aim
of monotonic regression is to fit a monotonic curve to the points (d_{ij}, δ_{ij}),
while making the sum of squared vertical deviations as small as possible (as
in least-squares linear regression). The point on the monotonic curve, $\hat{d}_{ij}$, is
the fitted or predicted value of d_{ij} from the monotonic regression. In judging
how good the fit is, we are now interested in how close the distances, d_{ij}, are
to the disparities, $\hat{d}_{ij}$, rather than the observed distances, δ_{ij}. This is because
we are only aiming to reproduce the rank order of the observed distances and
not the distances themselves. Hence, our measure of fit is obtained by cleverly
replacing δ_{ij} by $\hat{d}_{ij}$ in the formula for the stress ($\hat{d}_{ij}$ and δ_{ij} having the same
rank order). Thus in ordinal MDS, the stress is calculated as

$$\sqrt{\frac{\sum_{i<j}(d_{ij} - \hat{d}_{ij})^2}{\sum_{i<j} d_{ij}^2}}. \tag{3.2}$$

This is also known as Kruskal's stress, type I (which we shall refer to simply as
stress). The optimum configuration is determined by minimising this measure
of stress or some variant of it.

The points (δ_{ij}, d_{ij}) are shown by a cross in Figure 3.4. Note that while
the first and second points (counting from left to right) follow a monotonic
pattern, the third does not. To achieve monotonicity, the values of d_{ij} for the
second and third points are replaced by their mean. Similarly, the values of
d_{ij} for the fourth and fifth points are replaced by their mean. This leads to

the monotonic regression curve consisting of the series of solid lines shown in the plot. The vertical dotted lines represent the distances $d_{ij} - \hat{d}_{ij}$.

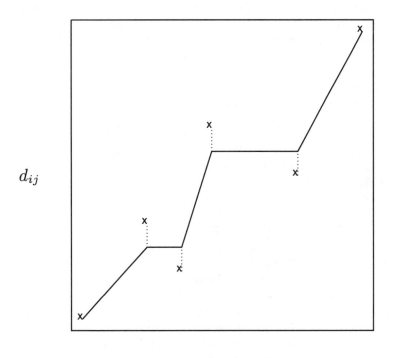

$$\delta_{ij}$$

Figure 3.4 *Example of monotonic regression*

Metrical scaling

Classical scaling could be described as metrical scaling since, in contrast to non-metrical scaling, the fitted and original distances are expressed in the same metric. However, the term metrical scaling usually seems to be reserved for something which may most naturally be thought of as related to non-metrical (ordinal) scaling in another way. In classical scaling, we supposed that the distances were Euclidean distances. In ordinal scaling, we made use only of the rank order of the δ_{ij}'s. This was tantamount to assuming that we had to make a monotonic transformation of the δ_{ij}'s to turn them into Euclidean distances. In metrical scaling, we assume that they can be transformed into Euclidean distances by some other parametric transformation. In some fields, there may be good reasons for supposing that such transformations exist, but we are not aware of any convincing arguments for introducing them in social science applications. However, we mention two special cases because they are closely linked to classical scaling. *Interval scaling* refers to the case where it is supposed that a linear transformation will turn the δ_{ij}'s into Euclidean distances. Instead of fitting a monotonic regression to the distances to obtain the disparities, we would now fit a linear regression. The disparities would then become the points on the regression line instead of points on the monotonic regression curve. The formula for stress remains the same except that the $\hat{d}_{ij}$'s would now be obtained from the least squares regression line. In the special case of *ratio scaling*, when the regression goes through the origin, we are back to the situation we faced in classical scaling because multiplying the δ_{ij}'s by a constant does not change the metric — if they were Euclidean before they will be Euclidean afterwards and vice versa. The difference here lies in the function which is being minimized. The Kruskal's stress formula applied in this case aims to achieve the closest degree of proportionality between the given distances and those fitted. Classical scaling aims to achieve the closest fit in a least squares sense. The two methods will often give very similar results and we shall use ratio scaling in one of the examples below.

3.4 Comments on computational procedures

Given the number of dimensions, k, the aim of MDS is to find a configuration in k dimensions such that the stress criterion used is minimised.

Most ordinal MDS computer packages start with an initial configuration in k dimensions, and then iteratively improve the configuration by moving the points short distances in such a manner as to reduce the stress slightly on each iteration. When further changes to the configuration do not reduce the stress (or not by more than some pre-specified tolerance level), the procedure ends and that configuration is the MDS solution. Typically, the method of steepest descent is used (Kruskal and Wish (1994), p.321-2, give the analogy of a blindfolded parachutist trying to find the lowest point in a terrain by following the gradient down hill).

Unfortunately, it is possible that a *local* minimum rather than the *global* minimum will be found. Repeating the process with different starting configurations to see whether the same minimum is found is one way of checking for this, but there is no absolute guarantee that there may not be some even smaller minimum lurking in a region of the space which has not been explored.

The MDS solution achieved depends on

i) the choice of initial configuration

ii) the stress criterion used

For example, the program PROXSCAL (available in SPSS v.10), with which many of the calculations in this chapter were done, arrives at a solution which minimises a stress function with d_{ij} replaced with δ_{ij} in the denominator of the formula for Kruskal's stress type I. There are other variants of stress which measure the differences between the distances and the disparities in slightly different ways.

Full discussion of such computational issues is outside the scope of this book, but the reader should be aware that different packages may give slightly different solutions. If the solutions are very different, this suggests that either there is no strong structure in the data, or that at least one of the solutions is a local rather than a global optimum, or that complete convergence has not been achieved for one or both solutions.

3.5 Assessing fit and choosing the number of dimensions

There are a number of ways of assessing the fit of a MDS solution. One method involves comparing the stress obtained for the solution with the guidelines shown in Table 3.3. These were developed by Kruskal (1964) and are based on empirical experience rather than theoretical criteria. These should always be used flexibly with an eye on the interpretability of the solution to which they lead.

Table 3.3 *Guidelines for assessing fit using stress*

Stress (Kruskal's type I)	Assessment of fit
0.20	poor
0.05	good
0.00	perfect

Another method that may be used to choose the number of dimensions is to examine a scree plot in which the stress is plotted against the number of dimensions. As the number of dimensions increases the stress decreases, but there is a trade-off between improving fit and reducing the interpretability of the solution. In the scree plot, we look for an "elbow" which is the point at which increasing the number of dimensions has little further effect on the stress. Again there is a strong subjective element in using this method, but experience shows that it often works well. See, for example, Figure 3.5 below.

There are also a number of useful diagnostic plots. In the case of ordinal scaling, the plots involve all pairs of δ_{ij}, d_{ij} and $\hat{d}_{ij}$, that may be examined to evaluate the fit of a MDS solution.

i) Plot of d_{ij} (the inter-point distance in the configuration) versus $\hat{d}_{ij}$ (the fitted value of d_{ij} obtained from the monotonic regression). If the MDS solution is a good fit, this plot should show a linear relationship with a 45 degree slope and only a small amount of scatter about the line. If little smoothing of the d_{ij}'s was necessary to produce the $\hat{d}_{ij}$'s, then they should be in almost the same rank order and close in value since they are measured on the same scale. See, for example, Figure 3.7.

ii) Plot of d_{ij} (the inter-point distance in the configuration) versus δ_{ij} (the observed distance or observation). If the solution is a good fit, d_{ij} and δ_{ij} should have approximately the same rank order and this plot should show a monotonic curve (either increasing or decreasing). See, for example, Figure 3.8.

iii) Plot of $\hat{d}_{ij}$ (the disparity or fitted value of the inter-point distance, d_{ij}) versus δ_{ij} (the observed distance or observation). The $\hat{d}_{ij}$'s are the "smoothed" versions of the d_{ij}'s constructed to have the same rank order as δ_{ij}. If a large amount of smoothing were required to achieve a monotonic curve (that is, the solution were a poor fit), this plot would show a large number of horizontal steps where the smoothing took place. See, for example, Figure 3.9.

For metrical scaling, the $\hat{d}_{ij}$'s are made to be proportional to the δ_{ij}'s. Therefore, the plots involving δ_{ij} are redundant, leaving only the plot of d_{ij} versus $\hat{d}_{ij}$ to be examined.

3.6 A worked example: dimensions of colour vision

We now illustrate these ideas and methods on an example which was originally analyzed by other means before the development of multidimensional scaling methods.

An experiment was conducted where subjects were asked to look at a screen which had two circular opaque glass windows. These windows were lit from two projectors behind the screen. Different colour filters could be inserted in the projectors. Fourteen colour filters were used, transmitting light of wave lengths $434m\mu$ to $674m\mu$. Each stimulus was combined with each other stimulus in a random order. The subjects were then asked to rate the degree of "qualitative similarity" between each pair of colour filters on a five-point scale. Further details, and the original analysis, will be found in Ekman (1954). The similarity matrix constructed by Ekman is given in Table 3.4. An ordinal MDS of these similarities was carried out.

Table 3.4 *Similarities between colours based on subjective judgements*

Colour	1	2	3	4	5	6	7	8	9	10	11	12	13	14
1	–	.86	.42	.42	.18	.06	.07	.04	.02	.07	.09	.12	.13	.16
2	.86	–	.50	.44	.22	.09	.07	.07	.02	.04	.07	.11	.13	.14
3	.42	.50	–	.81	.47	.17	.10	.08	.02	.01	.02	.01	.05	.03
4	.42	.44	.81	–	.54	.25	.10	.09	.02	.01	.01	.01	.02	.04
5	.18	.22	.47	.54	–	.61	.31	.26	.07	.02	.02	.01	.02	.01
6	.06	.09	.17	.25	.61	–	.62	.45	.14	.08	.02	.02	.02	.01
7	.07	.07	.10	.10	.31	.62	–	.73	.22	.14	.05	.02	.02	.01
8	.04	.07	.08	.09	.26	.45	.73	–	.33	.19	.04	.03	.02	.02
9	.02	.02	.02	.02	.07	.14	.22	.33	–	.58	.37	.27	.20	.23
10	.07	.04	.01	.01	.02	.08	.14	.19	.58	–	.74	.50	.41	.28
11	.09	.07	.02	.01	.02	.02	.05	.04	.37	.74	–	.76	.62	.55
12	.12	.11	.01	.01	.01	.02	.02	.03	.27	.50	.76	–	.85	.68
13	.13	.13	.05	.02	.02	.02	.02	.02	.20	.41	.62	.85	–	.76
14	.16	.14	.03	.04	.01	.01	.01	.02	.23	.28	.55	.68	.76	–

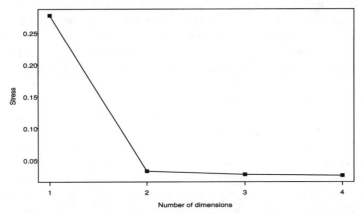

Figure 3.5 *Scree plot of stress by number of dimensions, colour data*

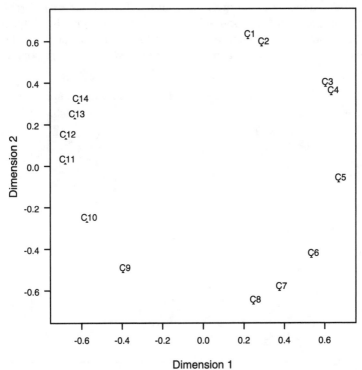

Figure 3.6 *Two-dimensional configuration plot from an ordinal MDS of colour data*

This is a case where we might guess in advance that a one-dimensional solution would be possible because the difference in wavelength between two colours is a continuous metric measuring how far apart the colours are. However, the scree plot given in Figure 3.5 shows that there is a big reduction in stress in passing from one to two dimensions, so there must be other factors which come into play when making subjective assessments of colour. The "elbow" at two dimensions indicates that there is little reduction in stress after two dimensions. Therefore, we select a two-dimensional solution. This solution has stress of 0.03 (3%) which according to Kruskal's guidelines is a good fit.

In the two-dimensional configuration (Figure 3.6), the points appear on a curve to give a "horseshoe" effect — a common phenomenon. At one extreme, are the violets (colours 1 and 2) and at the other are the reds (colours 11-14). As we go round the horseshoe, we encounter the colours in strict order of wavelength. However, it appears that subjects were making more subtle judgements in that reds are seen as closer to violets than to greens (colours 6-8), even though reds and greens are closer in terms of their wavelengths. Reference back to Table 3.4 confirms that this is not an accidental artefact of the MDS solution. There is clearly some other aspect of the perception of colour influencing the subject's comparisons than is conveyed by wavelength alone.

The three diagnostic plots are typical of what one finds with a reasonably good fit. On Figure 3.7, the points lie close to the 45 degree line; the curve in Figure 3.8 shows marked monotonicity, and Figure 3.9 has horizontal steps of short length reflecting the near monotonicity shown by the previous figure.

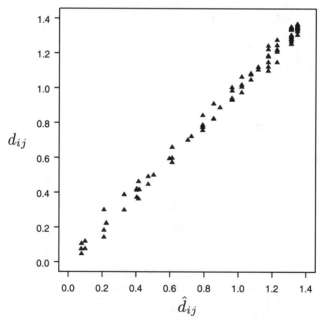

Figure 3.7 *Plot of d_{ij} versus $\hat{d}_{ij}$ from a two-dimensional ordinal MDS of colour data*

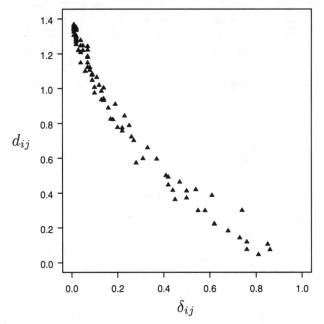

Figure 3.8 *Plot of d_{ij} versus δ_{ij} from a two-dimensional ordinal MDS of colour data*

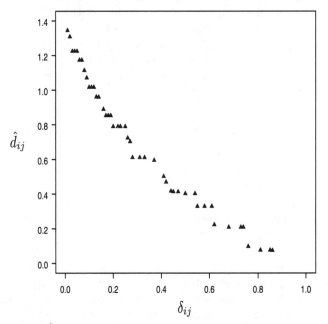

Figure 3.9 *Plot of $\hat{d}_{ij}$ of versus δ_{ij} from a two-dimensional ordinal MDS of colour data*

Before leaving this example, it is interesting to return to the one-dimensional solution plotted in Figure 3.10. The colours do not appear in order of their wavelength though there is a clear separation between the "blue" end of the spectrum (colours with low numbers) and the "red" (colours with high numbers). Within those two groups, however, there seems to be some inversion of the order one would have expected. The fit, of course, was not good in this case. The stress was 0.28, which on Kruskal's criterion, indicates a poor fit.

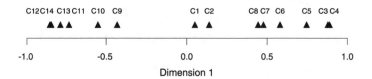

Figure 3.10 *MDS solution for colour data in one dimension*

The clear conclusion of our analysis is that colour perception involves more than is conveyed by the wavelength of the light. To return to the title of Ekman's paper, there appear to be two dimensions of colour vision.

3.7 Further examples and suggestions for further work

In this section, we give four further examples to illustrate the methods. We shall not carry out an exhaustive analysis on any of them, but focus on particularly interesting features which the individual examples show. You are invited to use these examples to explore the other options available in the various software packages. Two of the examples have already occurred in the chapter on cluster analysis, and our main interest in these cases will be to compare the two methods when applied to the same data.

Economic and demographic indicators for 25 countries

Table 3.5 shows the values of five economic and demographic indicators for a sample of 25 countries. The data refer to 1990 and they come from the United Nations Statistical Yearbook of 1997. The indicators are annual percentage population growth rate (Increase), life expectancy in years (Life), infant mortality rate per 1000 (IMR), total fertility rate (TFR), and Gross Domestic Product per capita in US dollars (GDP).

Ratio MDS was applied to these data. Since the data are in the form of a data matrix, the first stage of a MDS is to convert the data to a distance matrix showing the pairwise distances between countries. Since the variables differ greatly in terms of their variances, the variables are first standardized to have a variance of 1. Euclidean distances are then computed. Since we apply ratio MDS, the fitted distances will be proportional to the actual distances. You should try ordinal scaling and compare the results.

One aim of a MDS of these data might be to determine whether coun-

Table 3.5 *Economic and demographic indicators for 25 countries, 1990, UN Statistical Yearbook of 1997*

Country	Increase	Life	IMR	TFR	GDP
Albania	1.2	69.2	30	2.9	659.91
Argentina	1.2	68.6	24	2.8	4343.04
Australia	1.1	74.7	7	1.9	17529.98
Austria	1.0	73.0	7	1.5	20561.88
Benin	3.2	45.9	86	7.1	398.21
Bolivia	2.4	57.7	75	4.8	812.19
Brazil	1.5	64.0	58	2.9	3219.22
Cambodia	2.8	50.1	116	5.3	97.39
China	1.1	66.7	44	2.0	341.31
Colombia	1.7	66.4	37	2.7	1246.87
Croatia	−1.5	67.1	9	1.7	5400.66
El Salvador	2.2	63.9	46	4.0	988.58
France	0.4	73.0	7	1.7	21076.77
Greece	0.6	75.0	10	1.4	6501.23
Guatemala	2.9	62.4	48	5.4	831.81
Iran	2.3	67.0	36	5.0	9129.34
Italy	−0.2	74.2	8	1.3	19204.92
Malawi	3.3	45.0	143	7.2	229.01
Netherlands	0.7	74.4	7	1.6	18961.90
Pakistan	3.1	60.6	91	6.2	385.59
Papua New Guinea	1.9	55.2	68	5.1	839.03
Peru	1.7	64.1	64	3.4	1674.15
Romania	−0.5	66.6	23	1.5	1647.97
USA	1.1	72.5	9	2.1	21965.08
Zimbabwe	4.4	52.4	67	5.0	686.75

tries can be placed on a scale of development based on these five indicators. Therefore, the one-dimensional solution is of particular interest. Developed countries are generally characterised by low growth rate, high life expectancy, low infant mortality, low fertility and high GDP. If countries can be located on a single dimension of development, developed countries should be placed at one extreme with less developed countries (characterised by high growth rate, low life expectancy, high infant mortality, high fertility and low GDP) placed at the other extreme.

The stress (Kruskal type I) value for the one-dimensional solution was 0.17, suggesting a poor fit. The locations of the countries on a single dimension are given in Table 3.6. We find that the countries lie approximately where we would expect. At one extreme, we have the less developed mainly African and Asian countries, while at the other we have European countries, the USA and Australia. Since the one-dimensional fit is poor, however, you should go on to examine a two-dimensional solution to see whether a second dimension improves the fit and adds any new insight into the structure of the data.

Table 3.6 *Coordinate for each country from a one-dimensional ratio MDS of Economic and demographic indicators (arranged in increasing order)*

Country	Coordinate
Malawi	−2.027
Benin	−1.616
Cambodia	−1.414
Zimbabwe	−1.302
Pakistan	−1.133
Bolivia	−0.798
Papua New Guinea	−0.783
Guatemala	−0.706
El Salvador	−0.344
Peru	−0.277
Iran	−0.167
Brazil	−0.112
Colombia	0.036
China	0.188
Albania	0.220
Argentina	0.327
Romania	0.786
Greece	0.921
Australia	1.049
USA	1.105
Netherlands	1.158
Austria	1.164
Croatia	1.167
France	1.230
Italy	1.328

The stress value for the two-dimensional solution is 0.05, indicating a much better fit than the one-dimensional solution. Figure 3.11 shows the plot of d_{ij} versus $\hat{d}_{ij}$. The strong linear relationship between the distances in the configuration and the smoothed distances is a further indication that the data are well represented in two dimensions. As noted in Section 3.5, with ratio MDS the other two diagnostic plots, involving δ_{ij}, are redundant since δ_{ij} and $\hat{d}_{ij}$ have been made to be proportional.

The two-dimensional configuration is shown in Figure 3.12. The location of countries on dimension 1 is almost the same as in the one-dimensional solution. Dimension 1 could be interpreted as a measure of overall development. On the second dimension, Romania and Croatia stand out from the other countries. If you look at the profiles of these countries in Table 3.5, you can see that they both have characteristics associated with developed countries (low growth rate, moderately high life expectancy, fairly low infant mortality and very low fertility), which places them on the left hand side of the first dimension together with other developed countries. However, they have very low GDPs compared to other developed countries. Those countries located at the other end of the second dimension generally have high GDPs. Thus, the second dimension is largely a function of GDP.

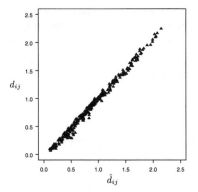

Figure 3.11 *Plot of d_{ij} versus $\hat{d}_{ij}$ from a two-dimensional ratio MDS of economic and demographic indicators*

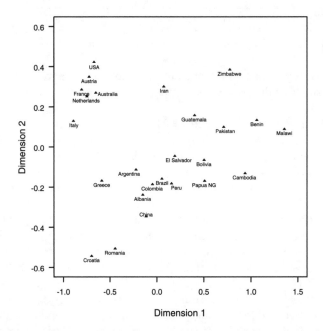

Figure 3.12 *Configuration of countries from a two-dimensional ratio MDS of economic and demographic indicators*

Persian archers

In Chapter 2, similarities between pairs of 24 archers (Table 2.13) were analyzed using cluster analysis. The data are described in Section 2.8. The similarities may also be analyzed using ordinal MDS.

From the scree plot in Figure 3.13, there is a suggestion of an elbow at two
dimensions, indicating that a two-dimensional solution may be adequate, but
three or four dimensions might improve the representation of the dissimilarities between the archers. The configuration for the two-dimensional solution
is plotted in Figure 3.14.

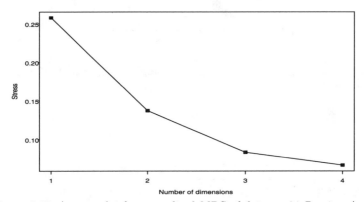

Figure 3.13 *A scree plot for an ordinal MDS of data on 24 Persian Archers*

Archaeologists want to know how the bas-reliefs were carved. Were they
the work of a single sculptor, several independent sculptors, or of one or more
teams of sculptors?

Figure 3.14 shows five archers (20 to 24) clustered together to the left of
centre near the bottom; eight archers (1-8) spread out upwards and slightly
diagonally on the left; the remaining archers (9 to 19) are spread out on the
right. Roaf (1983), p. 14-16, as we noted in Chapter 2, concluded that there
could have been three teams of sculptors. One working on the top section of
the staircase (1 to 8), another on the centre section (9 to 19) and a third on the
bottom section (20 to 24), these last five being so similar that they could be the
work of a single sculptor. Within this broad clustering into groups, adjacent
archers on the staircase tend to be close to each other in the configuration.

You may suggest explanations of why archers 1 to 8 are strung out in a
line in Figure 3.14, why archer 2 appears relatively close to archers 20 to
24, and why archer 12 is distant from the others. Then turn to Figure 3.15
where lines have been added joining points (archers) with similarity of 15 or
more. Such additions to the plot of the MDS solution can clarify whether the
relative positions of individual points in the configuration reflect their true
similarities. Points close together on the map but with low similarities will be
major contributors to the stress.

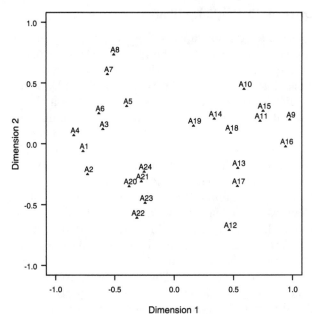

Figure 3.14 *24 Persian archers plotted in the two-dimensional space found through ordinal MDS*

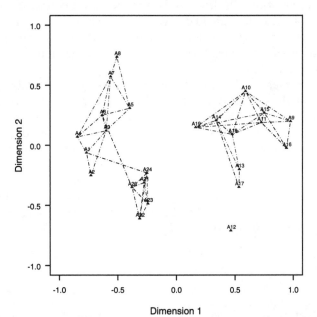

Figure 3.15 *24 Persian archers plotted in the two-dimensional space found through ordinal MDS, with lines drawn between pairs of archers with similarity $\geq$ 15*

Dialect words in 25 English villages

The data set on which this example is based was given in Section 2.5 where it formed the basis for demonstrating the various techniques of cluster analysis. We showed there that there was a fairly clear cluster structure in which the villages in each cluster were close together geographically as one would have expected. Given that the villages can be represented on a map in two dimensions, it is natural to ask whether one would obtain a similar map if linguistic similarity were used as a measure of distance. We would then be able to see whether the pattern of villages on the linguistic map was similar to their geographical situations. If this turned out to be the case, we would infer that the result of easier interchange between villages close together led to them having more words in common. But major topographical features, like rivers, roads and railways might make for greater similarity along the main lines of communication. There are no mountain barriers in that part of England, but a river like the Trent might well prove a barrier to easy communication.

Bearing these points in mind, you should carry out an ordinal MDS on the similarities in two dimensions. The stress is 0.14 which is not a particularly good fit in two dimensions, but given the particular interest of the two-dimensional plot in this case, it is given in Figure 3.16.

This should be compared with the map in Figure 2.6. The orientation is not the conventional one with north at the top of the diagram. The most northerly village on the map is V4 which occurs on the extreme right of the figure. The orientation will therefore be approximately correct if we rotate the figure anti-clockwise through 90 degrees. In that case, the Huntingdon village (V22) will be on the right, as for the conventional view. Rotating the figure has thus produced something fairly close to the map given in Chapter 2. This is shown in Figure 3.17.

A careful comparison of the "map" provided by your analysis with the true map will show a fairly good, but by no means exact, correspondence. This suggests that geographical factors play a major role in explaining the distribution of dialect words. It must be remembered, of course, that the measure of linguistic similarity we have used is based on a fairly small sample (60) of words.

In view of the relatively poor fit of the two-dimensional map, it is worth looking at the diagnostic plot of d_{ij} versus $\hat{d}_{ij}$. This is given in Figure 3.18. Although the fit is not as good as in some of the other examples, there is a broad correspondence between the d_{ij} and the $\hat{d}_{ij}$.

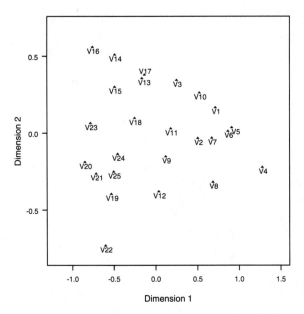

Figure 3.16 *Two-dimensional representation of 25 English villages*

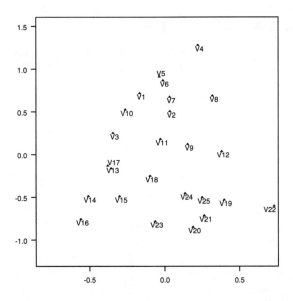

Figure 3.17 *The points on Figure 3.16 in a more conventional orientation*

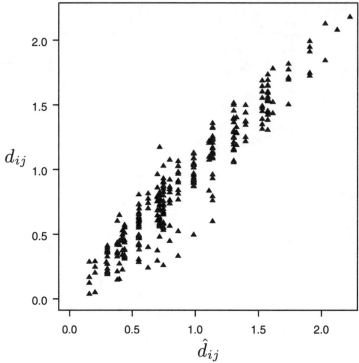

Figure 3.18 *Plot of d_{ij} versus $\hat{d}_{ij}$ for the dialect data*

Acoustic confusion of letters of the alphabet

In psychological experiments on memory, subjects may be asked to listen to and remember letters of the alphabet in some sequence. There is a risk that they may fail to give the right letters, not because of a failure of memory, but because they did not hear them clearly. Conrad (1964) reports the results of an experiment to investigate acoustic confusion in identifying letters of the alphabet. Three hundred post office employees wrote down the letters they thought they heard when letters were spoken against a background noise at a rate of one every five seconds.

Morgan (1973) calculated the similarities given in Table 3.7 by averaging the number of times the first letter was confused with the second, and the number of times the second was confused with the first, each letter being presented a total of 1440 times. The object of using MDS is to discover what led to letters being confused with each other.

Figure 3.19 shows the minimum values of Kruskal's stress type I for one-through six-dimensional solutions obtained from an ordinal MDS. There is unfortunately no clear elbow, and it is not until you come to the four-dimensional solution that the stress falls below 0.1. A two-dimensional solution will not be adequate, but that does not mean that it will be of no use.

Table 3.7 *Similarities between letters (average of number of times each was confused with the other), acoustic data*

	letter	w	g	c	q	p	t	b	d	e	u	v	h	f
1	w	*	6	6	8	8	10	35	27	18	30	21	18	13
2	g	6	*	41	142	185	128	182	151	242	222	172	5	3
3	c	6	41	*	73	385	274	203	90	129	78	81	22	8
4	q	8	142	73	*	446	265	137	106	118	153	61	32	0
5	p	8	185	385	446	*	786	237	235	283	95	125	27	13
6	t	10	128	274	265	786	*	227	201	287	40	72	32	18
7	b	35	182	203	137	237	227	*	322	379	139	290	18	8
8	d	27	151	90	106	235	201	322	*	418	101	252	17	5
9	e	18	242	129	118	283	287	379	418	*	190	174	53	15
10	u	30	222	78	153	95	40	139	101	190	*	426	28	6
11	v	21	172	81	61	125	72	290	252	174	426	*	18	4
12	h	18	5	22	32	27	32	18	17	53	28	18	*	81
13	f	13	3	8	0	13	18	8	5	15	6	4	81	*
14	s	7	7	20	10	7	15	4	9	23	8	4	194	824
15	x	3	3	11	3	7	7	3	5	25	15	1	191	483
16	l	38	6	2	7	6	2	9	6	11	3	3	16	41
17	j	13	20	16	19	26	14	35	24	10	31	25	23	13
18	k	21	5	11	20	45	19	13	12	16	25	23	43	37
19	m	25	25	18	10	33	15	21	16	72	28	12	18	35
20	n	39	34	26	12	29	20	23	27	112	35	31	55	40
21	a	83	39	11	11	16	20	27	28	26	38	26	104	19
22	o	77	27	5	9	13	40	14	10	27	25	20	50	28
23	i	22	13	8	13	13	15	15	14	114	55	9	4	7
24	r	9	10	3	10	1	10	3	5	19	5	6	8	18
25	y	16	12	12	5	4	9	8	7	12	8	11	4	12
26	z	97	5	8	14	34	10	26	14	10	12	21	53	121

	letter	s	x	l	j	k	m	n	a	o	i	r	y	z
1	w	7	3	38	13	21	25	39	83	77	22	9	16	97
2	g	7	3	6	20	5	25	34	39	27	13	10	12	5
3	c	20	11	2	16	11	18	26	11	5	8	3	12	8
4	q	10	3	7	19	20	10	12	11	9	13	10	5	14
5	p	7	7	6	26	45	33	29	16	13	13	1	4	34
6	t	15	7	2	14	19	15	20	20	40	15	10	9	10
7	b	4	3	9	35	13	21	23	27	14	15	3	8	26
8	d	9	5	6	24	12	16	27	28	10	14	5	7	14
9	e	23	25	11	10	16	72	112	26	27	114	19	12	10
10	u	8	15	3	31	25	28	35	38	25	55	5	8	12
11	v	4	1	3	25	23	12	31	26	20	9	6	11	21
12	h	194	191	16	23	43	18	55	104	50	4	8	4	53
13	f	824	483	41	13	37	35	40	19	28	7	18	12	121
14	s	*	575	60	40	41	44	49	42	44	24	20	15	120
15	x	575	*	13	8	15	11	15	9	7	5	11	6	78
16	l	60	13	*	74	68	115	76	203	101	86	193	123	47
17	j	40	8	74	*	222	46	106	161	87	14	18	118	150
18	k	41	15	68	222	*	82	144	246	101	13	27	31	80
19	m	44	11	115	46	82	*	846	151	339	83	65	69	48
20	n	49	15	76	106	144	846	*	360	89	77	52	58	58
21	a	42	9	203	161	246	151	360	*	594	20	36	28	26
22	o	44	7	101	87	101	339	89	594	*	54	56	22	53
23	i	24	5	86	14	13	83	77	20	54	*	292	164	7
24	r	20	11	193	18	27	65	52	36	56	292	*	194	30
25	y	15	6	123	118	31	69	58	28	22	164	194	*	41
26	z	120	78	47	150	80	48	58	26	53	7	30	41	*

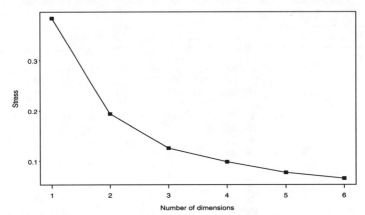

Figure 3.19 *A scree plot of stress against the number of dimensions used for the acoustic data*

The configuration of letters for the two-dimensional solution is shown in Figure 3.20.

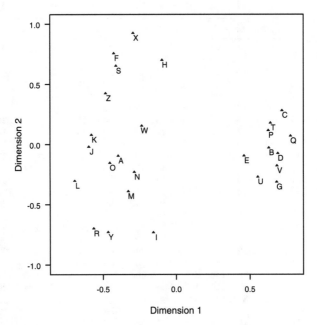

Figure 3.20 *Two-dimensional configuration of acoustic data from an ordinal MDS*

The tightest cluster consists of c, t, p, b, e, d, v, g, and q, u. Referring back to Table 3.7, you can see that these letters including q and u all do have relatively high similarities with each other. You can easily see that most members of this group share the "ee" sound, and it is presumably that fact which leads to them frequently being confused. It is not so obvious why q and u also come in this group, although q and u do have something in common.

This example shows that even when there is a rather poor fit, some meaning can still be extracted from the analysis. You may care to investigate solutions in three or more dimensions to see whether further meaningful groupings occur.

3.8 Further reading

Borg, I. and Groenen, P. (1997). *Modern Multidimensional Scaling, Theory and Applications*. New York: Springer-Verlag.

Cox, T. F. and Cox, M. A. A. (2001). *Multidimensional Scaling*. 2nd edition. London: Chapman and Hall/CRC.

Everitt, B. S. and Rabe-Hesketh, S. (1997). *The Analysis of Proximity Data*. London: Arnold.

Kruskal, J. B. and Wish, M. (1994). *Multidimensional Scaling*. In *Basic Measurement* Ed. M. S. Lewis-Beck. International Handbooks of Quantitative Applications in the Social Sciences, Volume 4. Sage Publications.

CHAPTER 4

Correspondence Analysis

4.1 Aims of correspondence analysis

Correspondence analysis (CORA) is an exploratory technique for analysing multi-way frequency tables, that is, cross-classifications of two or more categorical variables. We will focus on the analysis of two-way tables, but the analysis of multi-way tables depends on much the same set of ideas and will be discussed briefly later in the chapter. Like MDS, correspondence analysis aims to convert a table of numbers into a plot of points in a small number of dimensions — usually two. The term *correspondence analysis* derives from the French *Analyse Factorielle de Correspondences* which is the term used by Benzecri and others who developed the technique. However, the basic idea is found much earlier in attempts to scale the categories of contingency tables.

The usual way to begin the analysis of a two-way table would be to perform a chi-squared test of association between the row and column variables. If a significant association were found, the nature of the association could be explored by examining row and/or column percentages. However, when the number of categories is large, perhaps hundreds, comparisons of row (column) percentages across columns (rows) is difficult. The aim of CORA is to represent the raw data in a low-dimensional space so that it is easier to identify the key features of the data. CORA can be used to explore questions such as the following.

i) Are there row categories which have similar distributions over the column categories?

ii) Are any of the column categories similar with respect to their distributions over the row categories?

iii) Are the row/column categories ordered with respect to their distributions across the column/row categories? If so, are the categories fairly evenly spaced?

iv) Questions i) and ii) are concerned with the extent to which row/column distributions vary across the column/row categories. Further questions arise concerning the extent to which a cell given by a row and column category contributes to the overall association.

Form of data input

Frequency tables may arise in a number of ways. Most commonly, the row and column variables have mutually exclusive categories, in which case the table is called a *contingency table*. These variables may be nominal or ordinal. Indeed, as noted above, CORA may be used to explore whether a variable that is suspected to be ordinal may be treated as such. Table 4.1 shows a cross-tabulation of the political party voted for in the 1992 British general election by the main reason for voting for the chosen party (among those who voted). The data are from the British General Election Study (1992) (Heath, Jowell, Curtice, Brand, and Mitchell 1993). Both variables are nominal and have mutually exclusive categories.

Table 4.1 *Voting preference by reason, British General Election Study 1992*

Reason	Party Conservative	Labour	Lib Dem	Other	Refused	Total
Always vote that way	244	405	48	39	18	754
Best party	933	542	305	127	46	1953
My party had no chance	59	74	87	31	5	256
Total	1236	1021	440	197	89	2963

The categories of the row and/or column variables do not have to be mutually exclusive. For example, the data in Table 4.2 are from a survey on leisure activities in Norway (Clausen 1998). Respondents were asked whether they had engaged in any of ten activities in the previous year. Since each respondent may have engaged in more than one of the activities, the categories are not mutually exclusive. Other examples of this type of frequency table are found in market research where a number of product brands are rated on a series of attributes.

Table 4.2 *Leisure activities by occupation, Survey of Level of Living 1995, Norway*

Activity	Occupation Manual	Low NM*	High NM	Farmer	Student	Retired	Total
Sport event	301	497	208	50	254	187	1497
Cinema	261	550	250	27	339	157	1584
Dance/disco	361	534	204	59	324	216	1698
Cafe/restaurant	463	766	334	72	350	601	2586
Theatre	89	350	195	12	143	167	956
Classical concert	23	182	124	10	60	110	509
Pop concert	117	298	145	11	184	56	811
Art exhibition	104	379	219	21	152	213	1088
Library	130	352	153	17	272	264	1188
Church service	168	370	187	51	162	424	1362
Total	2017	4278	2019	330	2240	2395	13279

*NM denotes non-manual

4.2 Carrying out a correspondence analysis: a simple numerical example

To demonstrate how CORA is performed, consider a simple 3 × 3 table. As often happens in explaining techniques in multivariate analysis, it is easiest to grasp the idea if it is illustrated on a very simple example. We shall do that here, but it must be remembered that the full power of the technique can only be appreciated on much larger tables. In this particular case, it is to be expected that CORA will tell us little more than could be learnt from careful inspection of the table.

A cross-classification of attitude to abortion in the US and years of education is given in Table 4.3. The data are from the General Social Surveys of 1972-1974 and appear in Haberman (1978), p.264.

Table 4.3 *Attitude to abortion by education in the US, 1972-74: cell frequencies*

| | | Attitude | | | |
		Positive	Mixed	Negative	Total
Education	≤ 8	101	120	320	541
	9-12	599	341	756	1696
	≥ 13	475	161	308	944
	Total	1175	622	1384	3181

The chi-squared statistic for a test of independence between rows and columns is 157.58 on 4 degrees of freedom, which indicates that there is a significant association between education and attitude to abortion in the US.

Row profiles and row masses

To investigate this association further, we might look at the distribution of respondents across the attitude categories for each education category, that is the row proportions. We refer to the sets of row proportions as *row profiles*. The row profiles for the data in Table 4.3 are shown in Table 4.4. Also shown are the *row masses* (overall proportion in each row) and the *centroid* or *average row profile* (overall proportion in each column). We could also examine the distribution across education categories for each attitude category. To do so, we would calculate the column profiles, column masses and the average column profile. For now, we will consider only the row profiles but will discuss the role of column profiles later.

It is clear that there are marked differences in the three row profiles. The proportion with a positive attitude tends to increase as we move down the table; that is, attitude tends to become more positive for more years of education. However, note that we are only able to describe the pattern in these simple terms because the categories are ordered. In general, the categories will not be ordered and part of the purpose of the analysis will be to see whether

Table 4.4 *Attitude to abortion by education in the US, 1972-74: row profiles*

		Attitude			
		Positive	Mixed	Negative	Row mass
Education	≤ 8	0.187	0.222	0.591	0.170
	9-12	0.353	0.201	0.446	0.533
	≥ 13	0.503	0.171	0.326	0.297
Centroid or Average row profile		0.369	0.196	0.435	

there is an ordering which helps to make sense of the table. For our immediate purpose, we shall not make use of the ordering information.

For a table with three columns, the row profiles can be represented as points in two-dimensional space, because the proportions must add to 1. In Figure 4.1, they are represented by points inside an equilateral triangle, where the centre of the triangle corresponds to equal proportions of responses in each category, and a point nearer a vertex (a corner of the triangle) corresponds to a higher proportion in that category.

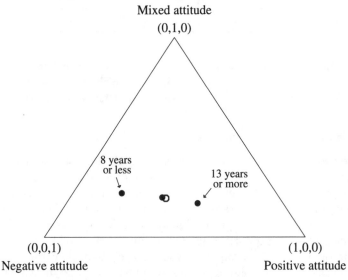

Figure 4.1 *The row profiles for the US education and attitude to abortion data. The open circle represents the centroid or average row profile and the solid dots represent the row profiles for the three educational groups.*

In Figure 4.1, the average row profile shows approximately equal proportions of positive and negative responses with a lower proportion of mixed responses. The profile for the highest education group (≥ 13 years) has a higher proportion of positive responses, while the profile for the lowest education group (≤ 8 years) has a higher proportion of negative responses (and a slightly higher proportion of mixed responses). The profile for the 9-12 year group is practically the same as that for the average row profile.

If there had been no association between attitude to abortion and education level, the row profiles would have been identical and the solid dots in Figure 4.1 would all have coincided with the average row profile (represented by the open circle). Their distance apart and their pattern therefore tell us something about the nature of the association. As a first step towards exploring this, we could calculate distances between each pair of row profiles, and between each row profile and the centroid. One possibility is to use the Euclidean distance, which is equal to the square root of the sum of squared differences between the profile values. For example, the Euclidean distance between the row profiles for ≤ 8 and 9-12 years of education is

$$\sqrt{(0.187 - 0.353)^2 + (0.222 - 0.201)^2 + (0.591 - 0.446)^2} = 0.221.$$

However, in CORA, each dimension is weighted inversely by the corresponding coordinate of the average row profile, so that column categories with a higher relative frequency do not dominate those with a lower relative frequency. The weighted Euclidean distance between the row profiles for ≤ 8 and 9-12 years of education is thus

$$\sqrt{\frac{(0.187 - 0.353)^2}{0.369} + \frac{(0.222 - 0.201)^2}{0.196} + \frac{(0.591 - 0.446)^2}{0.435}} = 0.354.$$

This measure of distance is often referred to as the chi-squared distance because weighting the squared difference between two profile values by the average profile value is analogous to weighting the squared difference between observed and expected values by the expected value. The full set of squared chi-squared distances is given in Table 4.5. In the last row of Table 4.5 are the squared chi-squared distances between row i and the average row profile or centroid, which we denote by d_i^2.

Table 4.5 *Squared chi-squared distances between row profiles, and between row profiles and the centroid, attitude to abortion by education, US, 1972-74*

	Row 1	Row 2	Row 3
1	0	–	–
2	0.125	0	–
3	0.445	0.099	0
Centroid	0.149	0.001	0.079

At this stage, we have a situation reminiscent of cluster analysis and MDS where the first step was to calculate a distance matrix. We could, indeed, proceed to carry out either kind of analysis on distance tables calculated from frequency tables. A cluster analysis, for example, might identify clusters of rows which had very similar profiles and this would suggest that those categories should be close together in some sense. MDS would provide a plot of points representing the rows and, again, it might be possible to infer something about the association from the pattern of the points. However, in CORA, our attention is focused more specifically on the nature of the association and the contribution which the various row and column categories make to it.

Inertia

In CORA, the term *inertia* is used to describe the measure of scatter or "variance" in the row (or column) profiles about the centroid. The total inertia is defined as

$$\sum_{i=1}^{I}(\text{mass for row } i) \times d_i^2$$

where I is the number of rows in the table. The term *inertia*, like the more familiar *degrees of freedom*, comes from mechanics; the analogy on which its use is based arises from the formula for inertia which is a mass multiplied by a squared distance.

From Tables 4.4 and 4.5, the total inertia for the US abortion data is

$$(0.170 \times 0.149) + (0.533 \times 0.001) + (0.297 \times 0.079) = 0.050.$$

It can be shown that the total inertia is related to the chi-squared statistic (X^2) divided by the grand total, n, that is

$$\text{Inertia} = \frac{X^2}{n}.$$

This result provides an interesting alternative way of looking at the chi-squared statistic. It now appears as a measure of the variation of the row profiles. If we interchange the rows and columns of the table the value of chi-squared remains the same, and it then follows that its value can also be represented as a measure of the variation of the column profiles.

The inertia measures the variation between rows which are multidimensional objects and which can thus be represented as points in space, as we have seen above. Equally, it measures variation between columns.

CORA depends on the fact that the inertia can be decomposed in another way, with each part measuring the variation in a single dimension. Because of the relationship between inertia and the chi-squared statistic, this is equivalent to decomposing the chi-squared statistic. If it turns out that most of the variation takes place in a small number of dimensions, two for example, it will be possible to picture the variation, and so perhaps interpret it in a meaningful way.

A two-dimensional representation

In the US education and attitude to abortion example, the row profiles may be represented in three-dimensional space because there are three rows and three columns. Actually, they may be represented perfectly in two dimensions, as in Figure 4.1, because any element in a row, say, may be obtained by subtracting the other two elements from the row total. However, CORA is most useful in much larger tables where the number of rows and columns is much greater than three. In such cases, it is desirable to reduce the dimensionality of the row profiles so that they may be plotted in two-dimensional, or at most three-dimensional, space. The question is how to find the coordinates of points representing the row profiles in two dimensions, and then to assess how good a representation of the original data these provide. In geometrical terms, the aim of CORA is to find a plane that is as close as possible to all the points and which also reproduces, as accurately as possible, the chi-squared distances between them. The row profiles are projected onto this plane to obtain points which represent the profiles in two dimensions.

Before describing how to do this, we look first at the interpretation of two-dimensional plots using the row profiles for the US education and attitude to abortion data as shown in Figure 4.2.

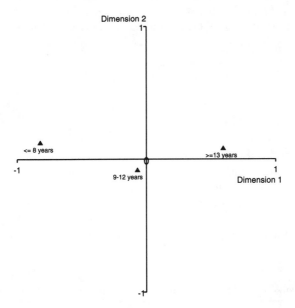

Figure 4.2 *Row profiles in two dimensions for attitude to abortion by education, US, 1972-74*

From Figure 4.2 we see that most of the variation between the row categories takes place in dimension 1 and, as we might expect, this dimension corresponds to "years of education". The row categories are thus given a metric with values

-0.82, -0.07 and 0.60 which are spaced at roughly equal intervals. However, there is a second dimension separating the middle group, 9-12 years, from the extremes. The variation in this dimension is much less.

If this kind of analysis is to be carried out for larger tables, we need some way of carrying out the decomposition in a manner which produces the successive dimensions algebraically. We now briefly outline the way that this is done, without going into the mathematical details which depend on what is known as the singular value decomposition of a matrix.

4.3 Carrying out a correspondence analysis: the general method

Pearson residuals

In correspondence analysis, the chi-squared statistic is partitioned in the manner described above. For the mathematics of the decomposition, it turns out to be more convenient to work, not with the profiles themselves, but with closely related quantities which we shall call *Pearson residuals*. We begin with an $I \times J$ matrix of observed frequencies, where I and J are the number of rows and columns respectively, and convert this to a matrix of Pearson residuals. These Pearson residuals are deviations between the observed frequencies and those expected under the model of independence. Denote the observed frequency for row i and column j of the table by O_{ij}, the total for row i by O_{i+} and the total for column j by O_{+j}. Denote the matrix of Pearson residuals by $\mathbf{C}$. The elements of $\mathbf{C}$ are

$$c_{ij} = \frac{O_{ij} - E_{ij}}{\sqrt{E_{ij}}} \ ,$$

where
$$E_{ij} = \frac{O_{i+}O_{+j}}{n} \qquad (i = 1, \ldots, I; \ j = 1, \ldots, J).$$

(Under the hypothesis of independence, each element of $\mathbf{C}$ has approximately the same distribution with zero mean and unit variance and thus the Pearson residuals are, in a sense, put on an equal footing.) If the row profiles are the same, the elements of each row of $\mathbf{C}$ will be zero. The size and pattern of the deviations from zero therefore tells us about the nature of the association. The matrix $\mathbf{C}$ for the US education and attitude to abortion data is given in Table 4.6.

Table 4.6 *Pearson residuals for attitude to abortion by education, US, 1972-74*

		Attitude		
		Positive	Mixed	Negative
Education	≤ 8	−6.99	1.38	5.52
	9-12	−1.10	0.51	0.67
	≥ 13	6.76	−1.74	−5.07

The singular value decomposition theorem tells us that we can write the typical element of $\mathbf{C}$ as

$$c_{ij} = \sum_{k=1}^{K} \lambda_k^{1/2} u_{ik} v_{jk} \qquad (i = 1, \ldots, I; \; j = 1, \ldots, J),$$

where K is the smaller of $I - 1$ and $J - 1$. The λ_k's are known as *eigenvalues* and their square roots, that is $\sqrt{\lambda_k}$ or $\lambda_k^{1/2}$, are the singular values. These are mathematical terms which are used here as convenient labels but, for our purposes, it is not necessary to know anything about their technical role in the derivation of the decomposition. The u_{ik}'s and v_{jk}'s may be thought of as scores attached to the rows and columns. In the simple analysis we carried out on the attitude to abortion data, we found scores for the row categories. In that case, $K=2$ so there was no dimensional reduction and there were two scores for each category. In the general case, each row category is represented by a point in K dimensions with coordinates $(u_{i1}, u_{i2}, \ldots, u_{iK})$, and the column categories by points with coordinates $(v_{j1}, v_{j2}, \ldots, v_{jK})$. Usually, however, we wish to represent row and column categories by points in a low-dimensional (preferably two) space.

The best approximation to c_{ij} in two dimensions is

$$c_{ij} \simeq \lambda_1^{1/2} u_{i1} v_{j1} + \lambda_2^{1/2} u_{i2} v_{j2} \, ,$$

where λ_1 and λ_2 are the two largest eigenvalues. Therefore, to present a graphical display in two dimensions, we could plot (u_{i1}, u_{i2}) and (v_{j1}, v_{j2}).

Usually, the coordinates are standardized in some way. We will use the following standardisation (implemented in SPSS v.10), where the standardized row coordinate for dimension k is calculated as

$$u_{ik}^* = \frac{u_{ik} \lambda_k^{1/4}}{\sqrt{O_{i+}/n}}.$$

The u_{ik}'s are multiplied by the inverse of the square roots of the row masses to ensure that row categories with high relative frequencies do not dominate rows with small frequencies. The u_{ik}'s are further multiplied by $\lambda^{1/4}$ so that more weight is attached to coordinates corresponding to the most important dimensions than to coordinates for less important dimensions.

The v_{jk}'s are transformed in a similar way to obtain standardized column coordinates, v_{jk}^*.

The eigenvalues of $\mathbf{C}$ are the principal way of judging the importance of the various dimensions. Associated with each dimension is an eigenvalue which represents the scatter of profiles about the centroid on that dimension, that is, the contribution to X^2 associated with that dimension. An alternative way of calculating the total contribution to X^2 is to take the sum of the eigenvalues. Eigenvalues may be compared across dimensions to assess the relative importance of each dimension in explaining X^2. The eigenvalues are ordered such that $\lambda_1 \geq \lambda_2 \geq \ldots \geq \lambda_K$. The dimensions are therefore constructed so that the first dimension explains the largest portion of X^2, or equivalently of the inertia, the second dimension explains the largest portion of the remaining inertia, and so on.

The proportion of X^2, or of inertia, explained by dimension k is

$$\frac{\lambda_k}{\lambda_1 + \lambda_2 + \ldots + \lambda_K}.$$

For the US education and attitude to abortion data, the proportion of inertia explained by the first dimension is $0.049/0.05 = 99\%$. The second dimension explains only 1% of inertia. Another way to look at the contribution of the first dimension is to estimate the value of X^2 if only the first dimension is considered. This is calculated as $\lambda_1 \times n = 0.049 \times 3181 = 151.0$ (compared with 157.58 for the full table). Most of the variation in the row profiles can therefore be expressed in one dimension.

Dual scaling

It is worth noting, at this point, a convergence between two different approaches to the analysis of association in contingency tables. In the approach which we have been following here, the category scores arise out of the decomposition of the table; but there was no thought at the outset of trying to assign scores to the categories. The other approach starts with what seems to be a quite different objective where the focus is on scaling both individuals and categories, and it sometimes goes under the name of dual scaling. We know a great deal about investigating the correlation structure of continuous variables. It might be possible to utilise this knowledge for contingency tables if there were some way of turning categorical data into continuous data. Dual scaling, also known as optimal scaling, asks whether there is some optimum way of assigning scores to individuals and categories so that the structure can be explored in terms of regression and correlation. We cannot go into the details here but the method turns out to be equivalent to correspondence analysis as we have described it. For some purposes, the scaling approach is the more natural way to approach some of the questions we outlined at the beginning of the chapter. For example, question iii) in Section 4.1 asked whether the categories were ordinal. If this were so, one might expect to be able to assign scores to the categories so that they formed an increasing (or decreasing) sequence. Our analysis shows that, in general, there can be no unambiguous answer to that question because the method assigns K sets of scores to the row categories and each will give a different ranking. However, as we saw with the abortion data, if one dimension is dominant we can reasonably treat the scores to which that leads as converting a nominal scale to an interval scale.

4.4 The biplot

The process by which row profiles are represented geometrically (Section 4.2) may be repeated for column profiles. It follows from the general representation of the singular value decomposition of $\mathbf{C}$ given above that the scatter of column profiles about the average column profile is equal to the scatter of row profiles about the average row profile. Therefore, the total inertia can be derived by considering either row profiles or column profiles. Also, the dimensionality of the row profiles is the same as that for the column profiles, even when the number of rows and columns are unequal. The maximum number of dimensions needed to represent either row profiles or column profiles is $K = \min(I - 1, J - 1)$. In terms of dimension reduction, the best-fitting line or the best-fitting plane to the row profiles explains the same proportion of inertia as the best-fitting line or the best-fitting plane to the column profiles.

Two-dimensional plots representing row or column profiles may be examined to identify whether any row or column categories have similar profiles. Row or column categories that have similar profiles will appear in close proximity on the plot. This could be useful to determine whether any row or column categories could be combined in subsequent analysis. However, also of interest, is how row and column categories interact with one another in contributing to the overall association. This aspect can be explored by the means of *biplots* which are plots of the points (u_{i1}^*, u_{i2}^*) and (v_{j1}^*, v_{j2}^*) on the same diagram. The purpose of a biplot can be seen by going back to the decomposition of the matrix of standardized residuals given above. This shows how the row and column score contribute to the overall size of the residual.

Recall that u_{ik}^* is the coordinate on dimension k representing row category i, and v_{jk}^* is the coordinate on dimension k representing column category j. The product $u_{ik}^* v_{jk}^*$ represents the joint contribution of row i and column j to the residual arising from dimension k. This is often spoken of as the "association" of column i and column j, but it must be distinguished from the overall association between the row categories and the column categories measured by X^2 with which we started the analysis. It is more accurately spoken of as a contribution to the overall association arising from a particular row and column.

In that sense, a large positive value for $u_{ik}^* v_{jk}^*$ indicates a positive relative association between row i and column j on dimension k. A large positive value is obtained if u_{ik}^* and v_{jk}^* are both large and positive or both large and negative, that is, the points for these categories appear close together on the biplot and far from zero on dimension k.

A large negative value for $u_{ik}^* v_{jk}^*$ indicates a negative relative association between row i and column j on dimension k. A large negative value is obtained if one of u_{ik}^* and v_{jk}^* is large and positive, and the other is large and negative, that is the points for these categories appear far apart on the biplot with neither point close to a value of zero on dimension k.

A value close to zero for $u_{ik}^* v_{jk}^*$ indicates no association between row i and column j on dimension k. A value close to zero is obtained if one or both of u_{ik}^* and v_{jk}^* is close to zero on dimension k.

There are two types of biplot that may be used in CORA: asymmetric plots and symmetric plots. It is the symmetric plot that we have been discussing above and will illustrate below. This is more generally useful.

Symmetric plots

Association between a row category and column category may be assessed according to the proximity of their profile points on the biplot. However, these proximities must be interpreted with caution. If the point for row category 1 is closest to the point for column category 2, we cannot say anything about the *magnitude* of their interaction in an absolute sense. We can only interpret it in *relative* terms. That is, we can say, for example, that individuals in row category 1 are relatively more likely (compared to the average row profile) to be in column category 2. It could be the case that, overall, there are very few individuals in column category 2; all we can say from the symmetric plot is that individuals in row category 1 are more likely to be in column category 2 than are individuals in the other row categories.

The coordinates of points in a symmetric biplot are scaled so that row or column points for rows or columns with high masses (marginal frequencies) do not dominate. They are further scaled (as described in Section 4.3) so that more weight is attached to coordinates corresponding to the most important dimensions than to coordinates for less important dimensions.

This procedure is illustrated on the biplot in Figure 4.3 showing the row and column profiles for the US abortion data. The coordinates for "≥ 13 years" (row category 3) and "positive" (column category 3) are both large and positive on dimension 1, giving a large positive value for $u_{31}^* v_{31}^*$. Thus, respondents with 13 or more years of education are relatively positively associated with having a positive attitude to abortion. The large negative coordinate for "≤ 8 years" (row category 1) on dimension 1 and the large positive coordinate for "positive" leads to a large negative value for $u_{11}^* v_{31}^*$. Respondents with 8 or fewer years of education are less likely to have a positive attitude than are more educated respondents.

Note that the coordinates for "9-12 years" and "mixed" on dimension 1 are both close to zero. Having 9-12 years of education is not associated with any attitude category, and having a mixed attitude is not associated with any education category. The latter can be seen by looking at the row profiles (Table 4.4). The proportion with a mixed attitude varies little across education categories.

Symmetric biplots are of fundamental importance for the interpretation of frequency tables and we shall therefore give two further examples. Biplots from CORA of the data in Table 4.1 and Table 4.2 are shown in Figure 4.4 and Figure 4.5 respectively.

From Figure 4.4, we can say that voting Labour is relatively associated with voting for a party because the respondent "always votes that way". We reach this conclusion because these particular row and column categories are in close proximity on the plot. Voting Conservative is relatively associated with voting for the "best party". Individuals who refused to state who they voted for are closer to Conservative voters in terms of their distribution across the main reason for voting categories. Conservative and Labour voters are relatively unlikely to vote for those parties because they thought their party had no chance of winning.

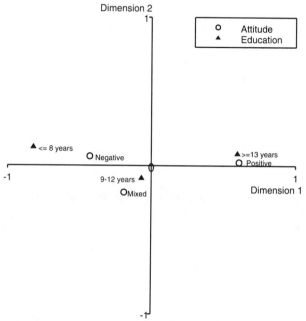

Figure 4.3 *Biplot for attitude to abortion by education, US, 1972-74*

It is important to emphasise the use of the word *relatively* in describing the association between a row category and a column category. From the type of plot shown above (a symmetric biplot), we cannot say anything about the *absolute* level of association. We can only say that a pair of row-column categories that are close together are more strongly associated than a pair of categories that are further apart.

CORA is more useful for analysing large contingency tables such as Table 4.2. From Figure 4.5, we can see, for example, that being retired is relatively associated with going to church, and students and low nonmanual workers are relatively more likely to go to the cinema and pop concerts than other occupation groups. Clausen (1998) also places an interpretation on the two dimensions. Dimension 1 separates the young (students) from the old (retired), while dimension 2 separates arts activities (e.g., going to a classical concert) from light entertainment (e.g., going to a disco).

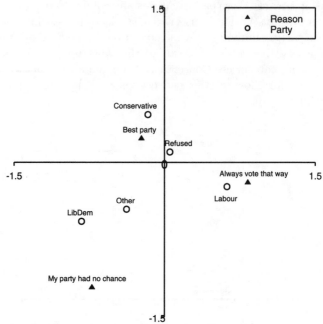

Figure 4.4 *Biplot for cross-tabulation of party preference by main reason for party choice, British General Election Study 1992*

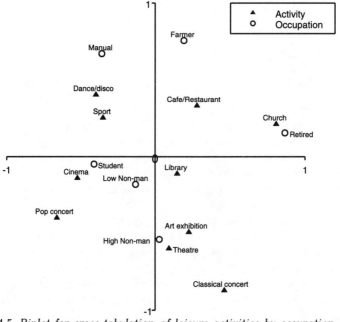

Figure 4.5 *Biplot for cross-tabulation of leisure activities by occupation, Survey of Level of Living 1995, Norway*

Asymmetric plots

In an asymmetric biplot, the row profiles are compared not with the column profiles but with the column vertices. (Alternatively, column profiles and row vertices are plotted simultaneously.) The point of doing this can be illustrated using the case of the 3×3 table. The column vertices are the following points in three-dimensional space: $(1,0,0)$, $(0,1,0)$ and $(0,0,1)$, where the first, second and third coordinates are the proportions in columns 1, 2, and 3 respectively. These points represent extreme cases in which all members of a row fall into one column category. The corner points of the triangular region shown in Figure 4.1 are the column vertices. Suppose that the row profile point for row 1 in the table was very close to the $(1,0,0)$ vertex, $(0.9, 0.04, 0.06)$. This would indicate that the majority of individuals in row 1 are in column 1. In other words, there is a large positive association between row category 1 and column category 1. On the other hand, suppose that the row profile point for row 1 was very close to one of the other vertices. This would indicate a high negative association between row category 1 and column category 1, since individuals in row category 1 are relatively unlikely to be in column category 1. For a higher dimension table, CORA is used to obtain an approximate lower dimensional representation of the row profiles and column vertices, preferably in one- or two-dimensional space. This low-dimensional plot is interpreted in the same way.

The main problem with the asymmetric map is that usually the row profiles are fairly close to the centroid (average profile). The addition of the column vertices to the plot alters the scale of the map so that the row profiles tend to appear very close together and are almost indistinguishable. However, if the scatter of the row profiles about the centroid is large (meaning the inertia is large), an asymmetric map may be useful. Also, there will be cases where it is unclear whether to view a cross-tabulation in terms of row profiles or column profiles; for example, if either the row or the column variables may be regarded as dependent variables. In this case, row and column profiles are of equal substantive interest and a symmetric plot may be more appropriate.

4.5 Interpretation of dimensions

It is sometimes possible to interpret or "label" the dimensions obtained from a CORA. We do so by examining the position of row/column categories along each dimension and thinking about what row/column categories that appear close together have in common, and what distinguishes those that appear far apart. For example, in his analysis of the Norwegian leisure data Clausen (1998) found that on the first dimension, light entertainment activities were grouped together and appeared far away from a cluster of arts activities. Biplots provide a visual display of such groupings of row/column categories. However, when interpreting a dimension, it is important to pay particular attention to those points which contribute the most to the inertia or scatter of points along that dimension.

Suppose we wish to interpret dimension k with respect to the row profiles. We can partition each point's contribution to the total inertia into its contributions to the inertia on each dimension in the CORA solution. The amount of inertia along dimension k explained by row point i is

$$\frac{(\text{mass for row } i) \times u_{ik}^{* \, 2}}{\sqrt{\lambda_k}} = u_{ik}^2.$$

Thus points corresponding to rows with a high row mass and with a large coordinate on dimension k will contribute the most to the inertia on dimension k. The amount of inertia explained by a given column point is calculated in a similar way. Points with relatively large contributions are most important to that dimension and provide the key to its interpretation. These values are examined together with the sign of the corresponding coordinates to interpret dimension k.

Table 4.7 *Coordinates and contribution to inertia of row points for attitude to abortion by education, US, 1972-74*

Education	Row mass	Coordinate		Contribution to inertia	
		Dim 1	Dim 2	Dim 1	Dim 2
≤ 8	0.170	−0.821	0.123	0.516	0.314
9-12	0.533	−0.069	−0.084	0.012	0.455
≥ 13	0.297	0.595	0.080	0.473	0.230

Table 4.8 *Coordinates and contribution to inertia of column points for attitude to abortion by education, US, 1972-74*

Attitude	Column mass	Coordinate		Contribution to inertia	
		Dim 1	Dim 2	Dim 1	Dim 2
Positive	0.369	0.606	0.022	0.609	0.022
Mixed	0.196	−0.191	−0.180	0.032	0.773
Negative	0.435	−0.428	0.062	0.359	0.206

We illustrate this process using the US education and attitude to abortion data. Table 4.7 shows the row masses and row coordinates in dimensions 1 and 2 (plotted in Figure 4.3), and the contribution of each row point to the inertia on each dimension in the two-dimensional solution. The contributions to inertia are expressed as proportions of the total inertia on that dimension. Table 4.8 shows the same quantities for the column categories. Starting with the row categories, we see that "≤ 8 years" and "≥ 13 years" explain similar proportions of the inertia on dimension 1. These two categories have coordinates which are opposite in sign. Thus we might label dimension 1 "level of education". Turning to the column categories (Table 4.8), we find that the "positive" category makes the largest contribution to inertia in dimension 1, followed by "negative". The corresponding coordinates in dimension 1 are opposite in sign, leading to this dimension being labelled "direction of attitude". Considering the interpretation of dimension 1 with respect to both row and column categories suggests that a high level of education is associated with a more positive attitude to abortion. In this case, we reached the same interpretation of the dimensions simply by examining the biplot in Figure 4.3. In general, however, row/column points with the largest coordinates will not always make the largest contributions to inertia since they may correspond to row/column categories with small relative frequencies. It is therefore important to examine both the coordinates of row/column points and the contributions of points to inertia in order to interpret dimensions.

4.6 Choosing the number of dimensions

As with MDS, the aim of CORA is to balance goodness-of-fit with parsimony when choosing the number of dimensions. The aim is to choose as few dimensions as possible, as this makes the task of interpretation easier. At the same time, the dimensions we choose to interpret should explain a reasonable amount of inertia. A commonly used tool is the scree plot, similar to the one used in MDS. The inertia for each dimension is plotted and the plot is examined for an "elbow", that is, the point after which there is little decrease in inertia. The scree plot for the Norwegian leisure activities data is shown in Figure 4.6. The maximum number of dimensions needed to represent the data is $\min(10\text{-}1, 6\text{-}1) = 5$. The elbow at three dimensions (or possibly four) suggests that two (possibly three) dimensions are sufficient to represent the data.

Another way of determining the number of dimensions is to examine the cumulative proportion of inertia explained by the dimensions. For example, in the case of the Norwegian data, the first two dimensions explain 90% of inertia, while the first three explain 99%.

The proportion of total inertia explained by the first k dimensions can be thought of as a measure of overall goodness-of-fit of the k-dimensional solution. We can also examine how well each row/column category is represented in k dimensions. Once again, we start by considering row categories. The total inertia of row point i is

$$\sum_{k=1}^{K} \lambda_k \times (\text{amount of inertia on dimension } k \text{ explained by point } i)$$
$$= \sum_{k=1}^{K} \sqrt{\lambda_k} \times (\text{mass for row } i) \times u_{ik}^{*2}.$$

The contribution of dimension k to the inertia of row point i is then

$$\frac{\sqrt{\lambda_k} \times (\text{mass for row } i) \times u_{ik}^{*2}}{\sum_{k=1}^{K} \sqrt{\lambda_k} \times (\text{mass for row } i) \times u_{ik}^{*2}}.$$

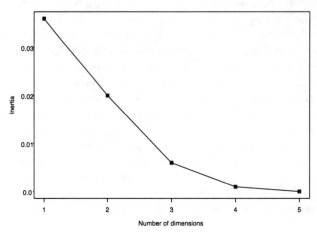

Figure 4.6 *Scree plot for Norwegian leisure activities data*

Table 4.9 *Contributions of dimensions to inertia of points for attitude to abortion by education, US, 1972-74*

		Dim 1	Dim 2	Total
Education	≤ 8	0.999	0.001	1.000
	9-12	0.949	0.051	1.000
	≥ 13	0.999	0.001	1.000
Attitude	Positive	1.000	0.000	1.000
	Mixed	0.968	0.032	1.000
	Negative	0.999	0.001	1.000

For example, using the information in Table 4.7 and the fact that the inertia on dimension 1 is 0.049, the total inertia of row point 1 (≤ 8 years of education) for the US abortion data is

$$(0.049 \times 0.516) + (0.001 \times 0.314) = 0.026,$$

and the contribution of dimension 1 to the inertia of row point 1 is

$$\frac{(0.049 \times 0.516)}{0.026} = 0.99.$$

The contribution of dimension k to the inertia of column points may be calculated in a similar way. The contributions of dimensions 1 and 2 to the inertia of row and column points for the US education and attitude to abortion data are shown in Table 4.9. These quantities measure how well each row and column point is described by each dimension. In this case, since dimension 1 is highly dominant, each row and column point is extremely well represented by the first dimension alone. Here, a maximum of two dimensions is required to represent the row/column profiles, so the sum of the contributions across dimension 1 and 2 equals 1. In general, however, when a larger table is analyzed the number of dimensions required to achieve a perfect fit will be large, and we will wish to assess the fit of a solution in considerably fewer dimensions.

4.7 Example: confidence in purchasing from European Community countries

We now illustrate the use of CORA with a more realistic example where the dimensions of the two-way cross-classification are large. The data are from the 1995 Eurobarometer Survey in which respondents from 15 countries in the European Community (EC) were asked about the confidence with which they would purchase various products or services from other EC countries (Reif and Marlier 1995). Using the responses to these questions, three binary items were constructed where 1 indicates confidence in buying a given product/service from another EC country and 0 indicates lack of confidence.

i) Confidence in buying food and/or wine

ii) Confidence in buying household electrical appliances

iii) Confidence in buying medical services and/or financial services

Since each item is binary, there are $2^3 = 8$ possible combinations of responses (response patterns) across the three items. A categorical variable was created with a category for each response pattern as described in Table 4.10.

Table 4.10 *Description of response patterns on three binary indicators of confidence in purchasing from other EC countries (1=Confident, 0=Not confident)*

Confident about purchasing ...	Response pattern
None of three types of product/service	000
Food/wine only	100
Household electrical only	010
Medical/financial services only	001
Any except medical/financial services	110
Any except household electrical appliances	101
Any except food/wine	011
Any of the products/services	111

A cross-tabulation of the response patterns on the three items by country of residence is shown in Table 4.11. The chi-squared statistic for this table is 3513.0 on 98 degrees of freedom, providing evidence of an association between confidence in shopping in the EC and country. To explore this association further, a CORA was carried out. We begin by examining the row profiles, that is the distribution of respondents across response patterns for each country (see Table 4.12). It can be seen that Spain and Greece have very similar row profiles. For example, respondents in these countries are the most likely to feel confident about purchasing any of the three types of product or service. Overall, there appears to be a division between Southern European countries (Italy, Greece, Portugal, and Spain) and Northern European countries. If respondents in Southern Europe feel confident about purchasing only one of the three products or services, it is relatively unlikely to be food/wine. In contrast relatively high proportions of respondents in Northern Europe (particularly Belgium, Germany, and the UK) feel confident about buying food/wine only.

Table 4.11 *Cross-tabulation of confidence in buying products/services from EC by country of residence, 1995*

Country	Response pattern for three products/services								Total
	000	100	010	001	110	101	011	111	
Austria	203	140	59	27	194	88	39	262	1012
Belgium	115	278	11	9	156	158	2	293	1022
Denmark	68	211	21	13	247	79	4	358	1001
Finland	115	177	44	20	209	85	29	363	1042
France	165	165	89	23	203	61	31	281	1018
Germany	243	600	46	48	272	380	27	483	2099
Greece	138	26	56	26	66	29	114	551	1006
Ireland	74	244	20	14	163	130	8	410	1063
Italy	144	81	118	117	118	102	141	318	1139
Luxembourg	76	51	10	8	30	40	6	155	376
Netherlands	51	155	13	10	145	243	12	385	1014
Portugal	281	63	103	41	85	44	56	379	1052
Spain	177	56	65	36	86	74	49	476	1019
Sweden	75	211	23	15	166	182	22	381	1075
UK	102	369	15	17	251	192	9	402	1357
Total	2027	2827	693	424	2391	1887	549	5497	16295

When a cross-tabulation has a large number of rows and columns (as in the case of Table 4.11), it can be difficult to pick out all of the important patterns in the data. CORA is particularly useful for analysing such tables. In the EC purchasing example, 7 dimensions are required to provide an exact representation of the row or column profiles (since the minimum of $(15-1)$ and $(8-1)$ equals 7). However, we hope that considerably fewer will be necessary to provide a good approximation. Table 4.13 shows the inertia on each dimension and the proportion of total inertia explained by each dimension. The first dimension is highly dominant accounting for almost 63% of the total inertia. The second and third dimensions each explain over 10% of inertia. We will consider the two-dimensional solution which accounts for 76.3% of

Table 4.12 *Row profiles for EC purchasing data*

Country	\multicolumn{8}{c}{Response pattern for three products/services}	Row mass
	000	100	010	001	110	101	011	111	
Austria	0.201	0.138	0.058	0.027	0.192	0.087	0.039	0.259	0.062
Belgium	0.113	0.272	0.011	0.009	0.153	0.155	0.002	0.287	0.063
Denmark	0.068	0.211	0.021	0.013	0.247	0.079	0.004	0.358	0.061
Finland	0.110	0.170	0.042	0.019	0.201	0.082	0.028	0.348	0.064
France	0.162	0.162	0.087	0.023	0.199	0.060	0.030	0.276	0.062
Germany	0.116	0.286	0.022	0.023	0.130	0.181	0.013	0.230	0.129
Greece	0.137	0.026	0.056	0.026	0.066	0.029	0.113	0.548	0.062
Ireland	0.070	0.230	0.019	0.013	0.153	0.122	0.008	0.386	0.065
Italy	0.126	0.071	0.104	0.103	0.104	0.090	0.124	0.279	0.070
Luxembourg	0.202	0.136	0.027	0.021	0.080	0.106	0.016	0.412	0.023
Netherlands	0.050	0.153	0.013	0.010	0.143	0.240	0.012	0.380	0.062
Portugal	0.267	0.060	0.098	0.039	0.081	0.042	0.053	0.360	0.065
Spain	0.174	0.055	0.064	0.035	0.084	0.073	0.048	0.467	0.063
Sweden	0.070	0.196	0.021	0.014	0.154	0.169	0.020	0.354	0.066
UK	0.075	0.272	0.011	0.013	0.185	0.141	0.007	0.296	0.083
Column mass	0.124	0.173	0.043	0.026	0.147	0.116	0.034	0.337	

inertia, although in practice it would be advisable to also examine the three-dimensional solution. We leave this as an exercise for the reader.

The coordinates on each dimension and the contribution to inertia of each dimension are shown for rows and columns in Table 4.14 and Table 4.15, respectively. To aid interpretation, countries and response patterns have been ordered according to their position on the first dimension. The countries with the largest contributions to the inertia on dimension 1 are (Italy, Greece, Portugal, Spain) with positive coordinates and (UK, Belgium, Germany) with negative coordinates (see Table 4.14). Thus, in general, dimension 1 contrasts Southern European countries with Northern European countries. Turning to the column points (Table 4.15), we find that the response patterns with large contributions to inertia on dimension 1 are (011,010,001) with positive coordinates and (100,101) with negative coordinates. The response patterns with large positive coordinates indicate confidence in purchasing a combination of one or two items, neither of which include food/wine; the categories with large negative coordinates correspond to confidence in purchasing selected items which include food/wine. Thus dimension 1 distinguishes between respondents who feel confident in purchasing food/wine from another EC country and those who do not.

We now consider the interpretation of dimension 2. Starting with row categories, we find that Greece (and to a lesser extent Spain) are contrasted with (Italy, France, Austria, Germany) (see Table 4.14). At this stage, it is not

Table 4.13 *Percentage of inertia explained by each dimension for the EC purchasing data*

		Inertia explained	
Dimension	Inertia	%	Cumulative %
1	0.135	62.6	62.6
2	0.030	13.7	76.3
3	0.026	11.8	88.1
4	0.016	7.4	95.5
5	0.006	2.9	98.4
6	0.002	1.1	99.4
7	0.001	0.6	100.0
Total	0.216		

Table 4.14 *Coordinates and contribution to inertia of row points for the EC purchasing data*

	Coordinate		Contribution to inertia	
Country	Dim 1	Dim 2	Dim 1	Dim 2
UK	−0.618	−0.033	0.087	0.001
Belgium	−0.596	−0.085	0.061	0.003
Germany	−0.552	−0.404	0.107	0.122
Netherlands	−0.472	0.504	0.038	0.092
Ireland	−0.405	0.326	0.029	0.040
Denmark	−0.402	0.198	0.027	0.014
Sweden	−0.365	0.256	0.024	0.025
Finland	−0.042	0.071	0.000	0.002
Luxembourg	0.160	0.305	0.002	0.012
France	0.202	−0.427	0.007	0.066
Austria	0.217	−0.427	0.008	0.066
Spain	0.657	0.436	0.073	0.069
Portugal	0.912	−0.227	0.146	0.019
Greece	1.032	0.887	0.179	0.282
Italy	1.057	−0.676	0.213	0.186

clear why countries should be grouped in this way, but we will return to these groupings after considering the interpretation of dimension 2 with respect to column categories. Of the column categories, dimension 2 is dominated by the group of respondents who feel confident in buying any of the three types of product or service (response pattern 111 with a contribution to inertia on di-

Table 4.15 *Coordinates and contribution to inertia of column points for the EC purchasing data*

Response pattern	Coordinate		Contribution to inertia	
	Dim 1	Dim 2	Dim 1	Dim 2
100	−0.759	−0.306	0.272	0.094
101	−0.617	−0.039	0.120	0.001
110	−0.327	−0.152	0.043	0.020
111	0.181	0.528	0.030	0.547
000	0.507	−0.334	0.087	0.081
001	1.021	−0.985	0.074	0.147
010	1.089	−0.664	0.137	0.109
011	1.607	−0.083	0.237	0.001

mension 2 of 0.547 — see Table 4.15). These respondents are contrasted with a group who would be confident about buying financial/medical services only (001) or household electrical goods only (010). This dimension may be loosely interpreted as a measure of the degree of confidence in purchasing goods or services from other EC countries. Response patterns located at the negative end of this dimension indicate confidence in buying selected products only, while the 111 pattern with a positive score indicates confidence in buying any product or service.

Next, we consider the association between row and column categories by considering the location of row and column points jointly. Figure 4.7 shows a symmetric biplot for the two-dimensional solution. The Southern European countries lie on one side of dimension 1, together with the response patterns which indicate lack of confidence in buying food/wine from other European countries; Northern European countries tend to lie at the other extreme of dimension 1, along with patterns indicating confidence in buying food/wine. This distinction may be partly explained by Spain, Portugal and Italy being wine-producing countries; respondents in these countries may be expressing the attitude that it is unnecessary to purchase wine from elsewhere. Dimension 2 has been interpreted as an indicator of the degree of confidence in purchasing from other EC countries. Located at the positive end of this dimension towards the 111 response pattern are Greece and Spain; Italy is located at the negative end of dimension 2 indicating less confidence in purchasing elsewhere.

To further evaluate the goodness-of-fit of the two-dimensional solution, we examine the contributions of dimensions 1 and 2 to the inertia of row and column points (Table 4.16). While many row and column categories are well represented in one dimension (e.g., Belgium, UK, 100 and 011), others are not (e.g., Finland, Luxembourg and 111). Finland and Luxembourg are not well represented in dimension 2 either, so their position in the two-dimensional solution relative to other countries and response patterns must be interpreted with caution. It is not surprising that some categories are not adequately represented in two dimensions since 24% of inertia remains unexplained by the first two dimensions. The addition of a third dimension may improve the representation of those categories which are poorly represented by two dimensions. In contrast, we find that response pattern 111 is well represented by a second dimension (also reflected in the high contribution of this column category to the inertia on dimension 2 — see Table 4.15).

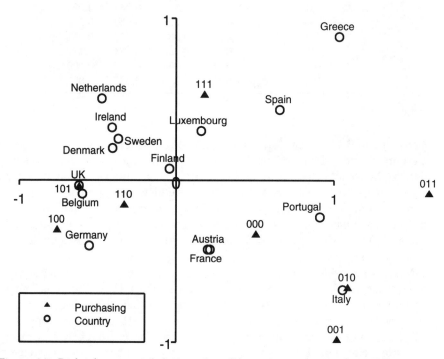

Figure 4.7 *Biplot for cross-tabulation of confidence in purchasing from EC country by country of residence*

Table 4.16 *Contributions of dimensions to inertia of points for the EC purchasing data*

		Contribution of dimension		
		Dim 1	Dim 2	Total
Country	Austria	0.174	0.314	0.489
	Belgium	0.914	0.009	0.922
	Denmark	0.375	0.042	0.417
	Finland	0.019	0.025	0.044
	France	0.128	0.268	0.395
	Germany	0.661	0.166	0.827
	Greece	0.697	0.241	0.938
	Ireland	0.671	0.203	0.875
	Italy	0.637	0.122	0.759
	Luxembourg	0.077	0.132	0.210
	Netherlands	0.357	0.190	0.547
	Portugal	0.751	0.022	0.773
	Spain	0.743	0.153	0.896
	Sweden	0.658	0.152	0.811
	UK	0.948	0.001	0.949
Purchasing	000	0.452	0.092	0.543
	100	0.872	0.066	0.938
	010	0.792	0.138	0.930
	001	0.504	0.220	0.724
	110	0.347	0.035	0.382
	101	0.579	0.001	0.581
	011	0.811	0.001	0.812
	111	0.198	0.790	0.988

4.8 Correspondence analysis of multi-way tables

So far, we have considered using CORA for the analysis of two-way tables. In one sense, therefore, we have been dealing only with the bivariate problem — one variable for the rows and one for the columns. Extensions to simple CORA, called multiple or joint CORA, have been developed for cross-classifications of more than two variables. All of them involve turning the multi-way table into a two-way table. We shall show, by means of an example, one way in which this may be done and then briefly outline a way of looking at the two-way table which generalises immediately to tables of any dimension.

A method where one variable is treated as dependent

As our example, we consider again the US education and attitude to abortion data; suppose we were interested in the association between attitude to abortion, education level and a third variable, religion. We could create a single new variable, called an *interactive* variable, with categories for each combination of education level and religion. Note that we could have combined education and attitude, or religion and attitude, in the same way to form a two-way table. However, it makes more substantive sense in this particular case to treat attitude as a dependent variable which is determined by combinations of the other variables. The new education/religion variable is cross-classified with attitude to abortion to create a two-way table (Table 4.17). The two-dimensional symmetric biplot from a CORA of these data is shown in Figure 4.8.

Table 4.17 *Attitude to abortion by education and religion, US, 1972-74*

| Education | Religion | Attitude | | | |
		Positive	Mixed	Negative	Total
≤ 8	Northern protestant	49	46	115	210
9-12	Northern protestant	293	140	277	710
≥ 13	Northern protestant	244	66	100	410
≤ 8	Southern protestant	27	34	117	178
9-12	Southern protestant	134	98	167	399
≥ 13	Southern protestant	138	38	73	249
≤ 8	Catholic	25	40	88	153
9-12	Catholic	172	103	312	587
≥ 13	Catholic	93	57	135	285
	Total	1175	622	1384	3181

Looking first at the scores derived for the row and column categories, we notice that attitudes move from "negative" to "positive" through "mixed" along dimension 1, whereas dimension 2 separates those with a mixed view from those who have a definite view one way or the other. For the education/religion categories, dimension 1 corresponds roughly to the "amount of education" variable with less than nine years on the left and 13 or more on the right. There is no such clear grouping for religion, suggesting that it is the interaction of education and religion rather than religion itself which counts.

This becomes clearer if we look at the biplot (Figure 4.8). We can see that protestants with ≥ 13 years of education have a relatively positive attitude to abortion, while catholics with the same level of education tend to have a negative attitude. Those with less education, of both religions, seem to have a more negative or mixed attitude.

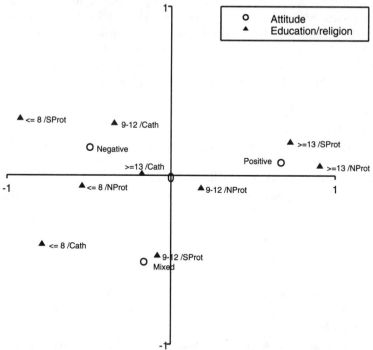

Figure 4.8 *Symmetric biplot from cross-tabulation of education-religion by attitude to abortion, US, 1972-74*

Direct analysis of the "data matrix"

The starting point of our discussion was the two-way contingency table which is very familiar from any elementary statistics course. However, there is much to be learnt if we go back a stage and see the table as the first step in summarising a data matrix. We introduced the idea of a data matrix in Chapter 1 and have shown in Chapters 2 and 3 how to construct distance matrices from it on the way to cluster analysis or MDS. The data expressed in a two- (or multi-) way contingency table can also be set out in a data matrix.

In the standard data matrix, the rows correspond to objects (individuals, sample members, etc.) and the columns to variables. The entry in any cell of the table is the value of the variable indicated at the head of the column for the object in the row. In the examples we have so far considered, the value of the variable was indicated by a single number — a 1 or 0 in the case of binary variables or a scale value for a metrical variable. For categorical variables, we use several numbers to indicate the category into which an individual falls. This is done by what is known as an indicator vector. Thus for example, if we consider the "religion" variable above listed in the order Northern Protestant, Southern Protestant and Catholic, the indicator vector would consist of 0s and 1s with 1 indicating the category into which the object falls. Thus a Southern

Protestant would be recorded as (010) and a Catholic as (001). This notation extends immediately to any number of categories; (001000) would indicate an object which falls into the third category of a six-category variable. Instead of having a single number in each cell of the data matrix, we therefore have a vector. (We could economise on notation and delete one element of the vector because if we know all elements but one, we can deduce what the other must be. This is exactly what we do if we have binary data where in the data matrix we only record whether or not the object is in the first category. This halves the size of the data matrix and means that we only have to deal with single numbers rather than vectors. That advantage is lost when we move to more that two categories, and it turns out to be more convenient to use the full indicator variables as we have defined them.)

A typical row of a data matrix for six variables might appear as follows:

$$100 : 01000 : 10 : 010 : 0001 : 01000$$

The colons are included only for clarity and the rows would normally be written without colons or spaces. Such a matrix is sometimes called the *super-indicator matrix*.

The data matrix, written in this fashion is a two-way array and can be subjected to correspondence analysis exactly like any other two-way table. The great merit of looking at the general problem in this way is that the two-way table is not now special in any way but can be included within the framework of a form of multiple correspondence analysis.

It is clear that an analysis carried out on such a data matrix will yield scores for both columns (categories of the variables) and rows (objects). The whole analysis can thus be viewed as a scaling exercise for both objects and categories.

We have described the data matrix as it arises when the categories are mutually exclusive. However, this representation may also be used quite generally. If an individual gives responses in more than one category, there will be a 1 entered in each column for which a response is given. In the "mutually exclusive case", the sum of every row is p, the number of variables. This is because there is precisely one 1 for each variable. In the general case, the row totals will be greater than p. In both cases, the column sums are the marginal totals for each category of each variable. For example, the first column refers to the first category of the first variable and its sum is simply the number of times an object falls into that category.

Apart from this property concerning the column sums, the data matrix looks very different to the contingency table, and it is not obvious that correspondence analysis of the two arrays will lead to essentially the same result. The fact that it does results from properties of the singular value decompositions of the two matrices which lie beyond the scope of this book.

4.9 Further examples and suggestions for further work

We give two further examples for you to work through, reminding you that we have already suggested some further analysis to be carried out on the example in Section 4.7. Both of the following examples demonstrate the usefulness of CORA as a way of identifying key patterns in fairly large two-way tables. For large tables such as these, a large number of dimensions is required in order to represent the data perfectly, but it turns out that in each case a two-dimensional solution provides a very good representation. We therefore focus on the interpretation of the first two dimensions.

As you work through these examples, you are encouraged to carry out some analyses of your own. For instance, you should assess the goodness-of-fit of the one- or two-dimensional solutions by examining scree plots and the contribution of each dimension to the inertia of row and column points. You might also examine three-dimensional solutions. Some suggestions for further analysis are given below.

Newspaper readership by occupation in the UK

Table 4.18 shows a cross-tabulation of newspaper readership by occupation in the UK. The data are from the 1999 Eurobarometer Survey (Melich, 1999). Respondents were asked which of the following newspapers they read most regularly:

1. Daily Mirror/Daily Record/Sunday Mirror

2. Sun/News of the World

3. Daily or Sunday Mail

4. Daily or Sunday Express/Sunday People

5. Times or Sunday Times

6. Daily or Sunday Telegraph

7. Guardian/Observer/Independent/Independent on Sunday

8. Other

9. None

Monday to Saturday and Sunday versions of the same newspaper have been grouped, as have two broadsheet newspapers (the Guardian and the Independent). For readers not familiar with British newspapers, we give a brief description here. First, a distinction can be made between tabloids and broadsheets. Newspapers in categories 1-4 above are tabloids, while 5-7 are broadsheets. A further distinction can be made based on political ideology. Newspapers 1, 2 and 7 are generally considered centre or left-of-centre, while 3, 4 and 6 are right-of-centre. The Times (5) is generally thought to be somewhere between the Guardian/Independent (7) and the Telegraph (6).

A chi-squared test provides evidence of an association between newspaper readership and occupation ($X^2 = 204.23$, degrees of freedom = 56, $p < 0.001$).

Table 4.18 *Cross-tabulation of newspaper read most regularly by occupation, UK, 1999*

Occupation	Newspaper									
	Mirr	Sun	Mail	Expr	Times	Teleg	Guard	Other	None	Total
Self-employed	8	16	12	6	6	4	7	2	6	67
Manager	9	14	19	8	17	10	18	12	9	116
Other white collar	20	20	20	9	7	10	6	12	8	112
Manual worker	90	103	34	28	10	4	7	30	28	334
Keeping house	51	64	19	9	9	4	6	24	29	215
Unemployed	22	30	6	1	2	3	5	12	11	92
Retired	60	50	52	31	8	19	7	48	31	306
Student	15	17	10	5	10	3	7	11	6	84
Total	275	314	172	97	69	57	63	151	128	1326

CORA may be carried out to obtain a low-dimensional representation of the two-way table. Seven dimensions are needed to provide a perfect representation. However, you will find that the first dimension explains 64.1% of the total inertia while the second explains a further 23.3%; therefore the first two dimensions explain 87.4% of inertia. There is a sharp decline in the inertia explained after the second dimension, suggesting that two dimensions should be adequate. As a further check of the goodness-of-fit of the two-dimensional solution, you should look at the contributions of each dimension to the inertia of the newspaper and occupation points.

Tables 4.19 and 4.20 show the coordinates and contribution to inertia from the two-dimensional solution for row (occupation) and column (newspaper) points respectively. To assist interpretation of the first dimension, occupations and newspapers have been ordered according to their location on dimension 1. You should begin with an interpretation of the first dimension. From Table 4.19, you will see that Manager (with a large positive coordinate) and Manual (with a large negative coordinate) make the largest contributions to inertia on dimension 1. Therefore, dimension 1 might be labelled "social class". Table 4.20 shows that the Times and Guardian/Independent (with large positive coordinates) and the Mirror and Sun (with large negative coordinates) make the largest contributions to inertia on dimension 1. The Telegraph also makes a moderate contribution to inertia and has a large, positive coordinate. Thus, this dimension distinguishes between left-of-centre tabloids and broadsheets. The occupation and newspaper categories are plotted on a biplot in Figure 4.9. On dimension 1, you can see that the broadsheets are located close to Manager; the Mirror and Sun are located close to Manual and Unemployed and, somewhat surprisingly, Keeping house.

Table 4.19 *Coordinates and contribution to inertia of row points for the UK newspaper data*

	Coordinate		Contribution to inertia	
Occupation	Dim 1	Dim 2	Dim 1	Dim 2
Manual	−0.508	−0.145	0.207	0.028
Keeping house	−0.412	−0.300	0.088	0.077
Unemployed	−0.403	−0.459	0.036	0.077
Retired	0.031	0.707	0.001	0.609
Other white collar	0.410	0.356	0.045	0.056
Student	0.468	−0.399	0.044	0.053
Self-employed	0.640	−0.339	0.066	0.031
Manager	1.359	−0.387	0.514	0.069

Table 4.20 *Coordinates and contribution to inertia of column points for the UK newspaper data*

	Coordinate		Contribution to inertia	
Newspaper	Dim 1	Dim 2	Dim 1	Dim 2
Mirror	−0.476	−0.021	0.150	0.000
Sun	−0.444	−0.387	0.148	0.188
None	−0.186	−0.039	0.011	0.001
Other	−0.017	0.400	0.000	0.096
Express	0.111	0.587	0.003	0.133
Mail	0.379	0.463	0.059	0.147
Telegraph	0.969	0.687	0.128	0.107
Times	1.160	−0.728	0.223	0.146
Guardian/Independent	1.356	−0.851	0.278	0.182

Among occupation categories (Table 4.19), dimension 2 is dominated by Retired which is quite distinct from the other categories, though closest to Other white collar. Turning to the newspapers (Table 4.20), a contrast can be seen between right-of-centre papers (Express, Mail, Telegraph) and more left-of-centre papers (Guardian/Independent, Times, Sun). This suggests that retired people are relatively more likely to read right-of-centre newspapers.

Some further observations can be made from the biplot in Figure 4.9. The self-employed and students have similar profiles with respect to their newspaper preferences, as do manual workers and the unemployed. Newspapers which have similar profiles across occupation categories are the Express and

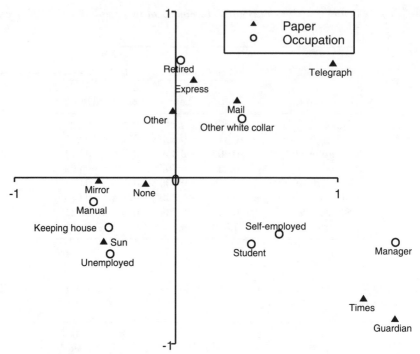

Figure 4.9 *Biplot for cross-tabulation of newspaper readership by occupation*

Mail (right-of-centre tabloids), Sun and Mirror (left-of-centre tabloids), and Times and Guardian/Independent (centre to left-of-centre broadsheets). The Telegraph appears quite distinct from the other newspapers.

Contraceptive method choice in Indonesia

Table 4.21 shows a cross-tabulation of choice of contraceptive method by age in Indonesia. The data are from the Indonesia Demographic and Health Survey of 1997 (CBSI 1998) and consist of 26833 women who were married at the time of the survey.

A chi-squared test provides very strong evidence of an association between contraceptive method and age (X^2=2644.98, degrees of freedom = 42, $p <$ 0.001). A correspondence analysis may be carried out to determine whether the patterns of contraceptive method choice across age groups may be represented in a small number of dimensions. Table 4.21 may be perfectly represented in six dimensions. However, you will find that the first dimension explains 73.0% of inertia and the first and second dimensions together explain 98.7%. This suggests that we should focus on the two-dimensional solution.

Tables 4.22 and 4.23 show the coordinates and contributions of points to the inertia of each dimension for the two-dimensional solution, for row (method) and column (age) points respectively. Starting with the interpretation of dimension 1 with respect to contraceptive method (Table 4.22), Sterilization and IUD (with negative scores) and Injectable (with a positive score) make the largest contribution to the inertia on this dimension. Dimension 1 contrasts clinical methods (Sterilization and IUD) with hormonal methods (Injectable, Implant, Pill). From Table 4.23, you can see that the first dimension also contrasts young women with older women. The clinical methods tend to be used for longer durations (or permanently in the case of sterilization) by older women who have achieved their desired number of children, while hormonal methods are typically seen as shorter-term methods and are predominantly used by younger women who wish to space rather than limit births.

Table 4.21 *Cross-tabulation of contraceptive method choice by age, Indonesia, 1997*

Method	Age (years)							Total
	15-19	20-24	25-29	30-34	35-39	40-44	45-49	
Pill	138	641	1009	902	852	491	219	4252
IUD	16	139	281	421	511	476	295	2139
Injectable	186	1037	1326	1126	837	398	133	5043
Implant	44	203	339	327	289	126	45	1373
Other modern	1	7	23	42	65	39	15	192
Sterilization	0	1	24	128	220	266	198	837
Traditional	7	64	142	205	231	185	124	958
None	674	1789	2262	1958	1928	1579	1849	12039
Total	1066	3881	5406	5109	4933	3560	2678	26833

Turning to the interpretation of the second dimension, the contraceptive method category "None" accounts for a large part of the inertia on this dimension. Notice also that "None" has the only negative coordinate on this dimension. Dimension 2 contrasts non-users (with a negative score) with users of any method (each method has a positive score). In Table 4.23, very young women (age 15-19) and the oldest age group are contrasted with women in their thirties and early forties. If you look at a biplot, you should find that the oldest women are located quite close to the "non-use" category on the second dimension since these categories have negative scores of a similar magnitude; reasons for high non-use among the 44-49 age group might include a low perceived risk of conception and conservative attitudes to family planning. The relatively high level of non-use for 15-19 year-olds could reflect early childbearing for women who marry at a very young age.

To assess further the fit of the one- and two-dimensional solutions, you should examine the contribution of each dimension to the inertia of the method and age group points.

Table 4.22 *Coordinates and contribution to inertia of row points for the Indonesia contraception data*

Method	Coordinate		Contribution to inertia	
	Dim 1	Dim 2	Dim 1	Dim 2
Pill	0.319	0.305	0.060	0.093
IUD	−0.692	0.571	0.142	0.163
Injectable	0.697	0.133	0.340	0.021
Implant	0.471	0.416	0.042	0.056
Other modern	−0.556	1.140	0.008	0.058
Sterilization	−1.660	0.519	0.320	0.053
Traditional	−0.540	0.582	0.039	0.076
None	−0.168	−0.413	0.047	0.481

Table 4.23 *Coordinates and contribution to inertia of column points for the Indonesia contraception data*

Age (years)	Coordinate		Contribution to inertia	
	Dim 1	Dim 2	Dim 1	Dim 2
15-19	0.229	−1.053	0.008	0.277
20-24	0.563	−0.316	0.171	0.091
25-29	0.484	−0.031	0.176	0.001
30-34	0.189	0.278	0.025	0.093
35-39	−0.161	0.394	0.018	0.180
40-44	−0.696	0.291	0.240	0.071
45-49	−0.953	−0.655	0.363	0.288

4.10 Further reading

Greenacre, M. and Blasius, J. (eds.) (1994). *Correspondence Analysis in the Social Sciences*. San Diego: Academic Press.

Michailidis, G. and De Leeuw, J. (1998). The Gifi system of descriptive multivariate analysis. *Statistical Science*, 13, 307-336.

CHAPTER 5

Principal Components Analysis

5.1 Introduction

The main aim of principal components analysis (PCA) is to replace p metrical correlated variables by a much smaller number of uncorrelated variables which contain most of the information in the original set. This greatly simplifies the task of understanding the structure of the data since it is much easier to interpret two or three uncorrelated variables than 20 or 30 that have a complicated pattern of interrelationships. In order to translate this objective into a practical method, we have to be more precise about what it is to retain "most of the information".

The central idea is based on the concept of the proportion of the *total variance* (the sum of the variances of the p original variables) that is accounted for by each of the new variables. PCA transforms the set of correlated variables $(x_1, \ldots, x_p)$ to a set of uncorrelated variables $(y_1, \ldots, y_p)$ called principal components, in such a way that y_1 explains the maximum possible of the total variance, y_2 the maximum possible of the remaining variance, and so on. The full set of p principal components fully explains the total variance:

$$\sum_{j=1}^{p} \operatorname{var}(y_j) = \sum_{i=1}^{p} \operatorname{var}(x_i).$$

However, if it turns out that the first few principal components account for a large enough part of the total variance, most of the variation in the x's being explained by the first few y's, then the remaining principal components can be discarded without too great a loss of information. It is usual to standardize the x's to unit variance before carrying out PCA so that each x-variable makes the same contribution to the total variance, and thus

$$\sum_{i=1}^{p} \operatorname{var}(x_i) = p.$$

Another aim of PCA is to interpret the underlying structure of the data in terms of the most important principal components. Often the principal components may be identified with some quantity of substantive interest. For example, the first principal component is frequently found to be positively correlated with each of the x's so that it can be interpreted as a measure of what is common to all the variables. Suppose we had measurements on the height, footsize, span, weight, waist and hip measurements of adult men. Because these six variables are all positively correlated with each other the first component, a measure of size, will be positively correlated with all of them. Sometimes components contrast one subset of the variables with another subset. Such components may be interpreted by thinking about what each subset of variables has in common. So, for the body measurements described above, the second component could be a contrast between the subset (height, footsize, span) and the subset (weight, waist, hip). The first component distinguishes between men by size, small men at one end and large at the other end of the scale. The second component has men who are relatively fat for their size at one extreme and men who are relatively thin at the other. The examples in Sections 5.5, 5.6 and 5.9 provide further illustrations of contrast effects and their interpretation.

PCA is analogous to correspondence analysis (CORA), but the two methods are applied to different types of data. PCA is a method for reducing the dimensionality of a set of correlated continuous variables, while CORA is a method for reducing the dimensionality of a cross-tabulation of associated categorical variables. In CORA, the dimensions are derived in order of importance in the sense that the first dimension explains the largest proportion of Pearson's chi-squared statistic or, equivalently, inertia. In PCA, the components are also derived in order of importance but in terms of the proportion of variance explained.

5.2 Some potential applications

i) Suppose we have examination scores in different subjects for a set of individuals. We would like to combine these in some way to obtain an overall measure of academic ability. From a PCA of the data, we find that the first component is positively correlated with each examination score. We interpret this component as a measure of general ability. However, the second component also explains a large amount of the variance in examination scores and contrasts science subjects with humanities subjects. From this, we conclude that ability may not be captured adequately by a single variable; there is a difference between ability in science subjects and ability in humanities subjects.

ii) Suppose we are interested in constructing a measure of deprivation. We have several indicators that could be thought of as deprivation measures. Each measures some slightly different aspect of deprivation and we somehow want to extract what is common to all of them to get at the core of the concept. What do we do? One approach would be to carry out a PCA on the set of indicators which would play the role of the x-variables. If the first component explains a large proportion of the total variation of the original deprivation measures, it could be used in place of the original variables as a measure of deprivation.

5.3 Illustration of PCA for two variables

Before dealing with PCA itself, we will give a simple example to show why transformations of the data might achieve the object we desire. Suppose we have observations on two variables, x_1 and x_2, which have the same scale and origin of measurement. Suppose also that they are highly correlated; this means that if we were to plot them they would lie close to the 45 degree line through the origin. Next we transform them to two new variables $y_1 = (x_1 + x_2)/\sqrt{2}$ and $y_2 = (x_2 - x_1)/\sqrt{2}$. Because x_1 and x_2 are highly correlated, it is clear that the second variable will have small values and will vary little; y_1 on the other hand, will have a much larger variance. The difference in the x's, therefore tells us very little about the variation between individuals whereas the sum tells us much more. In that sense, y_1 contains most of the information in the two variables about the variation among individuals. We now develop this idea into a full PCA using an example with two variables. Suppose that we have two x-variables, each having a variance 1, and that the correlation between them is 0.90. A scatterplot is shown in Figure 5.1.

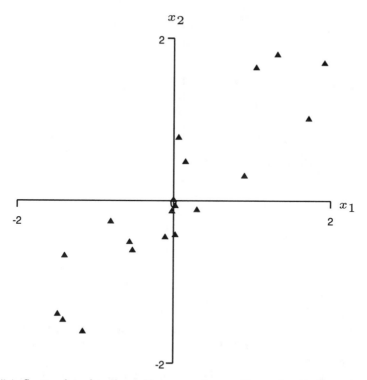

Figure 5.1 *Scatterplot of two variables, x_1 and x_2, with equal variance and correlation 0.90*

Finding the principal components for two variables involves an orthogonal rotation of the axes. An orthogonal rotation is one where the axes are kept at right angles to one another. The first principal component will be in the direction of greatest variance. This is found by minimizing the sum of the squared perpendicular distances from the observations to the first component. Once the first component is positioned, the second component is fixed since it must be orthogonal (at right angles) to the first. The dashed lines in Figure 5.2 represent the two principal components for the data in Figure 5.1.

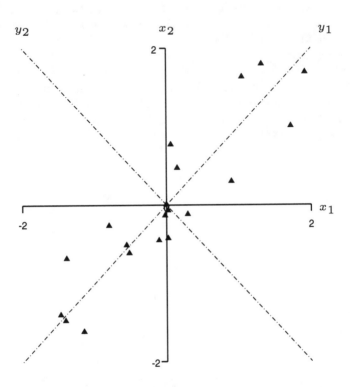

Figure 5.2 *The principal components (y_1 and y_2) for two correlated variables (x_1 and x_2) with equal variance*

If the variances of x_1 and x_2 are equal and the correlation between them is positive, as in Figure 5.2, then the first component will always lie at a 45 degree angle to the x_1 and x_2 axes. If x_1 and x_2 had unequal variance, the first component would lie closer to the axis with greater variance. Having found the principal components, we could plot our observations taking the components as our new axes. A scatterplot of the observations with the components as axes is shown in Figure 5.3.

Note that the proportion of total variance explained by the first component will depend on the degree of correlation between x_1 and x_2. The higher the correlation, the closer the observations will lie to the first component and the greater the proportion of variance explained by the first component.

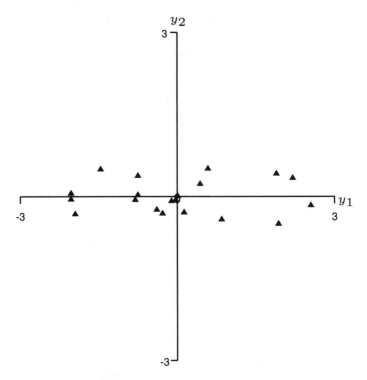

Figure 5.3 *Scatterplot of the data in Figure 5.1 taking the principal components as axes*

The variance of the first principal component is 1.90 and that of the second is 0.10. The original total variance of 2 is now apportioned unequally with the much greater part allocated to the first component.

Although it is less easy to visualize, we can imagine the same approach being applied with three x-variables. The scatter plot for three positively correlated variables such as height, footsize, and span will show a cluster of points with a shape similar to a slightly flattened rugby football. The first principal component will then lie along the major axis of the ball, the second — at right angles — along the largest diameter through the centre, and the third mutually at right angles to the first two.

In the example given above, the principal components could be found, approximately, by eye. This is sufficient to show that the answer we get will depend on the variances of the x's. In Figure 5.1, the variances were both 1 but we noted that it was easy to see what would happen if the variance of x_1, say, had been much greater than that of x_2. This would be equivalent to stretching the horizontal axis, the effect of which would be to reduce the slope of the line representing the first principal component. This makes the first component more like x_1. In general, the larger the variance of any variable, the more dominant its role. However, in many applications, especially in social sciences, the units of measurement are arbitrary and so any analysis whose results depend on these arbitrary scalings is practically meaningless. We can arrange that all the variables carry equal weight by first standardizing them so that they each have unit variance. In that case, the analysis depends only on the correlation matrix. Unless the original scale of the variables is meaningful, the first step in a PCA is, therefore, to compute the correlation matrix.

5.4 An outline of PCA

PCA transforms a set of correlated variables (x's) into a set of uncorrelated components (y's). The principal components are linear combinations of the x's which we write as:

$$
\begin{aligned}
y_1 &= a_{11}x_1 + a_{21}x_2 + \cdots + a_{p1}x_p \\
y_2 &= a_{12}x_1 + a_{22}x_2 + \cdots + a_{p2}x_p \\
&\vdots \\
y_p &= a_{1p}x_1 + a_{2p}x_2 + \cdots + a_{pp}x_p.
\end{aligned}
$$

Each component is a weighted sum of the x's, where the a_{ij}'s are the *weights*, or *coefficients*, for variable i on component j.

It is clear that there have to be some constraints on the a_{ij}'s. Otherwise we could make the variance of any y as large as we pleased simply by making the a_{ij}'s large enough. We can see what is required by going back to the treatment for two variables. We arrived at the principal components by rotating the axes while keeping them at right angles (orthogonal). In the general case, we have to find the equivalent algebraic formulation for orthogonal rotation. It turns out that this requires that the a_{ij}'s must satisfy the following conditions:

$$
\sum_{i=1}^{p} a_{ij}^2 = 1 \qquad (j = 1, 2, \ldots, p),
$$

and

$$
\sum_{i=1}^{p} a_{ij} a_{ik} = 0 \qquad (j \neq k; \ j = 1, \ldots, p; \ k = 1, \ldots, p).
$$

Another way of describing what these conditions do is to say that they leave the relative positions or the configuration of the points unchanged.

An important consequence of the orthogonality condition is that, as stated in Section 5.1, the total variance of the y's is equal to the total variance of the x's, that is

$$
\sum_{j=1}^{p} \mathrm{var}(y_j) = \sum_{i=1}^{p} \mathrm{var}(x_i).
$$

This means that the total variance does not change; rather variance is re-distributed among the variables. The y's are derived in decreasing order of importance such that y_1 has maximum variance and, therefore, explains the largest proportion of the total variance. The first principal component may thus be thought of as the best one-dimensional summary of the data. The second component, y_2, is derived so that it has the second largest variance, subject to the constraints

$$\sum_{i=1}^{p} a_{i2}^2 = 1 \quad \text{and} \quad \sum_{i=1}^{p} a_{i1}a_{i2} = 0,$$

so that it is orthogonal to (uncorrelated with) y_1. The first two components provide the best two-dimensional summary of the data. Subsequent components are derived in decreasing order of variance, each component being uncorrelated with the previous components.

The mathematical problem that this procedure poses, therefore, is to find a method of determining the a_{ij}'s so that the components have the required properties. Although this seems a formidable problem, it is easily solved because it turns out to be equivalent to a well known problem in matrix algebra concerned with finding what are called the eigenvalues and eigenvectors of a matrix — the matrix in this case being either the covariance matrix or, more commonly, the correlation matrix. There are standard algorithms which determine the weights a_{ij} and the variances of the principal components. The latter are usually denoted by $(\lambda_1, \lambda_2, \ldots, \lambda_p)$ and are listed from largest to smallest.

Choosing the number of principal components

As for CORA and MDS, one aim of PCA is to be able to plot the data in one, two or three dimensions without losing too much information. When choosing the number of components, the aim is to retain as small a set as possible but, at the same time, to have a sufficient number to provide a good representation of the original data. The variance of component j is the eigenvalue λ_j. Since the components are derived in order of variance, $\lambda_1 \geq \lambda_2 \cdots \geq \lambda_p$. If the x's are standardized so that the correlation matrix is analyzed, the sum of the variances of the x's will be equal to p. Thus the sum of the eigenvalues, the total variance of the y's, will also equal p.

The proportion of total variance explained by component j is

$$\frac{\lambda_j}{\lambda_1 + \lambda_2 + \cdots + \lambda_p}.$$

The proportion explained by the first k components together is

$$\frac{\lambda_1 + \lambda_2 + \cdots + \lambda_k}{\lambda_1 + \lambda_2 + \cdots + \lambda_p}.$$

In practice, these proportions are often expressed as percentages.

There are a number of criteria that may be used to decide how many components should be retained:

i) Retain the first k components which explain a "large" proportion of the total variation, say 70-80%.

ii) If the correlation matrix is analyzed, retain only those components with eigenvalues greater than 1. The logic behind this rule of thumb is that a component with an eigenvalue of 1 explains the same amount of variation as one of the original x's. However, Jolliffe (1972), suggests that retaining components with eigenvalues greater than 0.7 is better than the cut-off at 1.

iii) Examine a scree plot. This is a plot of the eigenvalues versus the component number. The idea is to look for the "elbow" which corresponds to the point after which the eigenvalues decrease more slowly. Adding components after this point explains relatively little more of the variance. See Figure 5.4 for an example of a scree plot.

iv) Consider whether the component has a sensible and useful interpretation.

Interpretation

The weight given to variable i on component j is a_{ij}. The relative sizes of the a_{ij}'s reflect the relative contributions made by each variable to the component. To interpret a component, we examine the pattern in the a_{ij} values for that component.

Often, the coefficients are rescaled so that coefficients for the most important components (i.e., the ones that explain the most variation) are larger than those for less important components. These rescaled coefficients, called *component loadings*, are the coefficients for reconstructing the x's from the y's as explained in Section 5.8 and are calculated as

$$a_{ij}^* = \sqrt{\lambda_j} a_{ij} \qquad (i = 1, \ldots, p; \ j = 1, \ldots, p).$$

When the correlation matrix of the x's is analyzed, a_{ij}^* may be interpreted as the correlation coefficient between variable i and component j. This is especially useful for interpretation.

5.5 Examples

Children's personality traits

To demonstrate the interpretation of components, consider the following example. In a study of children's personality traits, a sample of children were scored on the eight variables listed in Table 5.1 which also gives the loadings, a_{ij}^*, for the first two components of the correlation matrix. The first two components explain 77% of the total variance of the scores.

For the first component, the loadings are all fairly large, positive, and of a similar magnitude. Each variable is positively correlated with the first component. The first component might therefore be interpreted as some overall measure of personality: a child who scores highly on each trait would have a

Table 5.1 *Component loadings for the first two principal components, children's personality trait data*

Personality trait	Variable	a_{i1}^*	a_{i2}^*
Mannerliness	x_1	0.68	0.58
Approval seeking	x_2	0.60	0.59
Initiative	x_3	0.65	−0.52
Guilt	x_4	0.65	−0.59
Sociability	x_5	0.61	0.57
Creativity	x_6	0.71	−0.61
Adult role	x_7	0.69	−0.49
Cooperativeness	x_8	0.67	0.61
Proportion of total variance		44%	33%

high score on this component, while a child who has a low score on each trait would have a low score on this component.

The coefficients for the second component have a bipolar structure. Mannerliness (x_1), approval seeking (x_2), sociability (x_5), and cooperativeness (x_8) all have relatively high positive loadings, while the other traits have relatively high negative loadings. To interpret this component, we need to think about what the variables in each subset have in common, and how they differ from the other subset of variables. The variables with positive loadings measure how well a child relates to other people. The variables with negative loadings are more internal to the individual and might be thought of as measures of the child's independence. Therefore, the second component may be interpreted as a contrast between these two aspects of personality. Children with a high score on this component will tend to get along relatively well with others but to show relatively less independent behaviour. In contrast, children with a low score on this component will tend to be relatively more independent in their behaviour but not to be so sociable. Children who are located towards the middle of the second component will have roughly equal scores for (x_1, x_2, x_5, x_8) as for (x_3, x_4, x_6, x_7).

The above structure is quite common in a PCA of a correlation matrix. If the original variables, the x's, are all positively correlated with one another, the loadings for the first component will have the same sign (either positive or negative) and will tend to be of a similar magnitude. Subsequent components will be contrasts between subsets of variables.

Subject marks

Table 5.2 shows the pairwise correlation coefficients between subject scores for a sample of 220 boys. The data are taken from Lawley and Maxwell (1971), p. 66.

It is always good practice to inspect the correlation matrix before embark-

Table 5.2 *Pairwise correlation coefficients between subject marks*

	Gaelic	English	History	Arithmetic	Algebra	Geometry
Gaelic	1.00					
English	0.44	1.00				
History	0.41	0.35	1.00			
Arithmetic	0.29	0.35	0.16	1.00		
Algebra	0.33	0.32	0.19	0.59	1.00	
Geometry	0.25	0.33	0.18	0.47	0.46	1.00

ing on any analysis. This may reveal anomalous entries and enables us to check the plausibility of the data. From such an inspection, one may be able to anticipate the results of a PCA because the analysis only makes explicit what is already implicit in the correlations. In this case, each subject score is positively correlated with each of the scores in the other subjects, indicating that there is a general tendency for those who do well in one subject to do well in others. The highest correlations are between the three mathematical subjects and to a slightly lesser extent, between the three humanities subjects suggesting that there is more in common within each of these two groups than between them.

A PCA was carried out on this correlation matrix and produced the eigenvalues shown in Table 5.3. Only the first two eigenvalues are greater than 1. Also shown are the percentage of variance explained by each component and the cumulative percentage of variation explained by the first k components. For example, the first two components explain just over 64% of the total variation.

Table 5.3 *Variance explained by each principal component, subject marks data*

Component	Variance explained		
	Variance (Eigenvalue)	%	Cumulative %
1	2.73	45.48	45.48
2	1.13	18.81	64.29
3	0.62	10.26	74.55
4	0.60	10.05	84.59
5	0.52	8.71	93.30
6	0.40	6.70	100.00

Figure 5.4 shows the scree plot. There is an elbow at the third component. The third and subsequent components have similar eigenvalues which means that they each explain a similar but small proportion of the total variance.

From Table 5.3 and Figure 5.4, we would conclude that the first two components should provide an adequate representation of the x's.

The loadings for the first two components are shown in Table 5.4 and are plotted in Figure 5.5. The loadings of the first component are all positive and fairly large. Remembering that they can be interpreted as correlations between the subject scores and the component, we infer that the first component represents something which is common to performance on all subjects. This could be what we usually describe as general academic ability. The second component is a contrast between humanities subjects and mathematical subjects. Boys who are better at humanities subjects than at mathematics will score highly on this component and conversely. This component measures the contrast between the two types of ability. Figure 5.5 makes the same point graphically with the three mathematics loadings appearing as a tight cluster separated from those for the humanities on the vertical dimension.

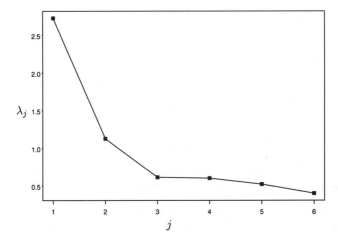

Figure 5.4 *Scree plot showing eigenvalue by number of principal component, subject marks data*

Table 5.4 *Loadings for the first two principal components, subject marks data*

Subject	a_{i1}^*	a_{i2}^*
Gaelic	0.66	0.44
English	0.69	0.29
History	0.52	0.64
Arithmetic	0.74	−0.42
Algebra	0.74	−0.37
Geometry	0.68	−0.35

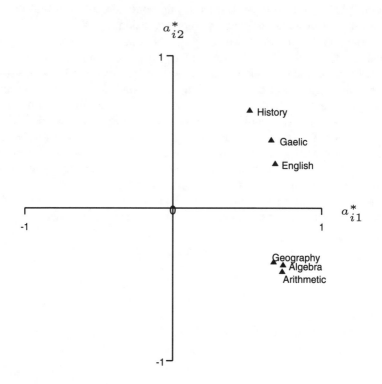

Figure 5.5 *Plot of loadings for the first two components, subject marks data*

Employment in 26 European countries

The percentage of the total workforce employed during 1979 in each of the nine industries listed below was recorded for 26 European countries (see Hand, Daly, Lunn, McConway, and Ostrowski 1994).

x_1: Agriculture   % employed in agriculture
x_2: Mining   % employed in mining
x_3: Manufacture   % employed in manufacturing
x_4: Power   % employed in power supply industries
x_5: Construction % employed in construction
x_6: Service   % employed in service industries
x_7: Finance   % employed in finance
x_8: Social   % employed in social and personal services
x_9: Transport   % employed in transport and communications

There are very large differences in the variances of these variables (ranging from 0.14 to 241.70) and, as we wish to give the variables equal weight, the correlation matrix will be analyzed. The lower part of the correlation matrix is shown in Table 5.5. From this, we can see that the percentage employed in agriculture has a fairly strong and negative correlation with the percentage employed in all other industries apart from mining. The percentage employed in manufacturing is quite highly, positively correlated with the percentage in construction. Also, the percentage employed in social and personal services is positively correlated with the percentages in service industries and the percentage in transport and communications. This matrix does not conform to the pattern of positive correlations observed in the previous examples and so we would not expect the first principal component to have all positive loadings.

Table 5.5 *Correlation matrix for European employment data*

	x_1	x_2	x_3	x_4	x_5	x_6	x_7	x_8	x_9
x_1	1.00								
x_2	0.04	1.00							
x_3	−0.67	0.45	1.00						
x_4	−0.40	0.41	0.39	1.00					
x_5	−0.54	−0.03	0.49	0.06	1.00				
x_6	−0.74	−0.40	0.20	0.20	0.36	1.00			
x_7	−0.22	−0.44	−0.16	0.11	0.02	0.37	1.00		
x_8	−0.75	−0.28	0.15	0.13	0.16	0.57	0.11	1.00	
x_9	−0.57	0.16	0.35	0.38	0.39	0.19	−0.25	0.57	1.00

The percentage explained by each of the nine principal components is given in Table 5.6. The first two components account for just over 60% of the total variation. The first three components have eigenvalues greater than 1, but the eigenvalue for component 4 is very close to 1. The scree plot (Figure 5.6) shows some suggestion of an elbow at component 3, indicating that between two and four components might be considered. The loadings for the first three components are shown in Table 5.7.

Table 5.6 *Variance explained by each component, European employment data*

Component	Variance explained		
	Variance	%	Cumulative %
1	3.49	38.75	38.75
2	2.13	23.67	62.42
3	1.10	12.21	74.63
4	1.00	11.05	85.68
5	0.54	6.04	91.71
6	0.38	4.26	95.97
7	0.23	2.51	98.48
8	0.14	1.52	100.00
9	0.00	0.00	100.00

Notice that the first component is almost perfectly negatively correlated with agriculture (x_1), and has low correlations with mining (x_2) and finance

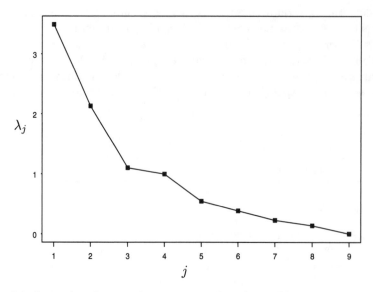

Figure 5.6 *Scree plot of eigenvalue versus number of component, European employment data*

Table 5.7 *Loadings for first three components, European employment data*

Industry	a_{i1}^*	a_{i2}^*	a_{i3}^*
Agriculture	−0.98	0.08	−0.05
Mining	0.00	0.90	0.21
Manufacture	0.65	0.52	0.16
Power	0.48	0.38	0.59
Construction	0.61	0.07	−0.16
Service	0.71	−0.51	0.12
Finance	0.14	−0.66	0.62
Social and personal services	0.72	−0.32	−0.33
Transport	0.69	0.30	−0.39

(x_7). The other variables have moderate to high positive correlations with the first component. The first component may be interpreted as distinguishing countries with agricultural economies from those with industrial economies. The second component is positively correlated with mining (x_2), manufacturing (x_3), power (x_4), and transport (x_9) and is negatively correlated with service (x_6), finance (x_7) and social and personal services (x_8). This component may be interpreted as a contrast between countries with relatively large and relatively small service sectors. These contrasts are clearly brought out in Figure 5.7. There is no obvious interpretation of the third component, which confirms our hope based on the scree plot that two components would be adequate.

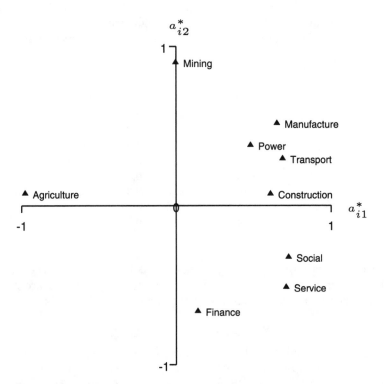

Figure 5.7 *Plot of loadings for first two components, European employment data*

5.6 Component scores

Suppose we wish to compute an individual's score on a particular component. From the PCA, we have

$$y_j = a_{1j}x_1 + a_{2j}x_2 + \cdots + a_{pj}x_p,$$

where $x_1, x_2, \ldots, x_p$ are all standardized to have a mean of zero and a variance of 1, and where y_j has variance λ_j. See, for example, the scores for y_1 and y_2 plotted in Figure 5.3.

However, it is more usual to standardize the component scores to have unit variance, so that $\tilde{y}_j = y_j/\sqrt{\lambda_j}$ has variance 1. Hence,

$$\tilde{y}_j = \tilde{a}_{1j}x_1 + \tilde{a}_{2j}x_2 + \cdots + \tilde{a}_{pj}x_p,$$

where

$$\tilde{a}_{ij} = \frac{a_{ij}}{\sqrt{\lambda_j}}.$$

These $\tilde{a}_{ij}$'s are referred to as the *component score coefficients*.

For example, since in our PCA of the subject marks data we have interpreted the first component as a measure of overall ability, a boy's ability could be estimated by computing component scores, $\tilde{y}_1$, using the coefficients in the first column of Table 5.8. Using the coefficients in the second column, we would be able to score individuals on the humanities versus mathematics dimension, $\tilde{y}_2$.

Table 5.8 *Component score coefficients for first two components, subject marks data*

Subject	$\tilde{a}_{i1}$	$\tilde{a}_{i2}$
Gaelic	0.24	0.39
English	0.25	0.26
History	0.19	0.57
Arithmetic	0.27	-0.37
Algebra	0.27	-0.33
Geometry	0.25	-0.31

The coefficients for computing standardized scores ($\tilde{y}_j$) on the first two components from the PCA of the European employment data are given in Table 5.9. For example, a country's score on component 1 would be calculated as

$$\tilde{y}_1 = -0.28x_1^* + \cdots + 0.20x_9^*,$$

where x_1^* is the standardized version of x_1, the percentage employed in agriculture, etc. The scores for the first two components have been calculated for each of the 26 countries. Figure 5.8 shows a scatterplot of the scores on the first component versus the scores on the second component. On the first component, Turkey and, to a lesser extent, Yugoslavia, stand out from the other countries. This would suggest that Turkey and Yugoslavia had more agricultural economies than the other countries in 1979. The second component separates the capitalist Western countries (negative scores) from the communist Eastern countries (positive scores). Since countries with high percentages in service industries (service, finance and social) would have a low score on this component, this suggests that the capitalist Western countries had a larger service sector than the communist Eastern countries.

Table 5.9 *Component score coefficients for the first two components, European employment data*

Industry	$\tilde{a}_{i1}$	$\tilde{a}_{i2}$
Agriculture	−0.28	0.04
Mining	−0.00	0.42
Manufacture	0.19	0.24
Power	0.14	0.18
Construction	0.17	0.04
Service	0.20	−0.24
Finance	0.04	−0.31
Social and personal services	0.21	−0.15
Transport	0.20	0.14

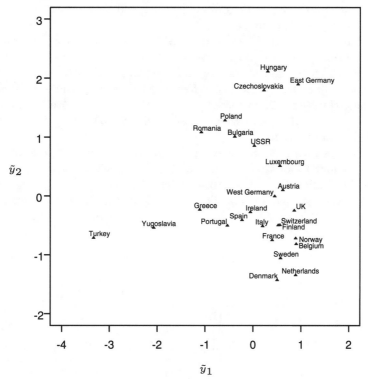

Figure 5.8 *Twenty six European countries plotted using standardized principal component scores, European employment data*

5.7 The link between PCA and multidimensional scaling, and between PCA and correspondence analysis

The link between PCA and multidimensional scaling

A basic PCA only requires the correlation matrix, but in order to calculate the component scores for the n individuals (for example, the 26 European countries), it is necessary also to know the values of the p (standardized) variables given in the $n \times p$ data matrix $\mathbf{X}$ as defined in Chapter 1:

$$
\mathbf{X} = \begin{pmatrix}
x_{11} & x_{12} & \cdots & x_{1p} \\
x_{21} & x_{22} & \cdots & x_{2p} \\
\vdots & & & \vdots \\
x_{n1} & x_{n2} & \cdots & x_{np}.
\end{pmatrix}
$$

Multidimensional scaling can be carried out using just the distances between individuals — but it gives no information about the variables from which the distances might have been constructed. Given the data matrix, either one or the other or both techniques could be applied to produce a plot of the individuals in, say, two dimensions.

Standard metrical MDS, like ordinal MDS, uses an iterative procedure to find a solution that minimises some stress criterion. PCA uses an algebraic procedure to maximise the variances of the components. In general, the results will be different (although maybe not very different where the structure is strong). However, where the distances have been calculated from a data matrix, *classical* MDS (sometimes called *principal coordinate analysis*) gives the PCA solution exactly.

We demonstrate their relationship below by carrying out a PCA of the correlation matrix for the economic and demographic data for 25 countries analyzed in Section 3.7. Table 5.10 shows the proportion of variance explained by each of the five components. The first component is dominant, explaining 80% of the total variance.

Table 5.10 *Variance explained by each component, economic and demographic data*

Component	Variance explained		
	Variance	%	Cumulative %
1	4.01	80.28	80.28
2	0.57	11.38	91.66
3	0.25	5.05	96.71
4	0.09	1.92	98.62
5	0.06	1.38	100.00

Table 5.11 shows the loadings for the principal components. The first component is highly correlated with all the variables, negatively with x_1, x_3, and x_4 and positively with x_2 and x_5. A country with a high rate of population

increase, low life expectancy, high infant mortality rate, high fertility rate and low GDP would have a low score on this component. A country with a low rate of population increase, high life expectancy, low infant mortality rate, low fertility rate, and high GDP would have a high score. Therefore, the first component may be interpreted as a measure of overall development. The second component is positively correlated with GDP and, to a lesser degree, rate of population increase and fertility rate. Life expectancy and infant mortality rate do not contribute to this component. Countries with a low rate of population increase, low fertility rate, and low GDP would have a low score on this component.

Table 5.11 *Loadings for all five components, economic and demographic country data*

Index	a_{i1}^*	a_{i2}^*	a_{i3}^*	a_{i4}^*	a_{i5}^*
Population increase	−0.86	0.39	0.32	0.09	−0.06
Life expectancy	0.95	0.03	0.24	0.05	0.19
Infant mortality rate	−0.95	−0.01	−0.21	0.20	0.13
Fertility rate	−0.95	0.19	−0.01	−0.22	0.12
GDP	0.76	0.62	−0.21	0.01	0.00

The standardized scores for each of the 25 countries on the first two components have been plotted in Figure 5.9. Note the close resemblance between this plot and the two-dimensional metrical MDS configuration for these data (see Figure 3.12).

The link between PCA and correspondence analysis

The usual derivation of PCA starts with the correlation matrix. An alternative derivation, giving exactly the same results, involves the singular value decomposition (SVD) of the standardized data matrix: $\mathbf{X}^* = \{x_{ti}^*\}$, where x_{ti}^*, the value for individual (row) t on variable (column) i, has been standardized to zero mean and unit standard deviation. This parallels the SVD of the matrix of Pearson residuals in CORA described in Section 4.3.

From the SVD, we obtain u_{tj}'s (corresponding to rows or individuals) and v_{ij}'s (corresponding to columns or variables) and singular values, $\sqrt{\lambda_j}$'s, such that

$$x_{ti}^* = \sum_j \sqrt{\lambda_j} u_{tj} v_{ij}.$$

It turns out that the square of the singular value is λ_j, the jth eigenvalue of the correlation matrix (and the variance explained by the jth principal component); u_{tj} is $\tilde{y}_j$, the score for individual t on principal component j; and $\sqrt{\lambda_j} v_{ij}$ is the component score coefficient of the variable i on principal component j. As for CORA, it is possible to plot individuals (rows) and variables (columns) in a single bi-plot, using the first two PCs.

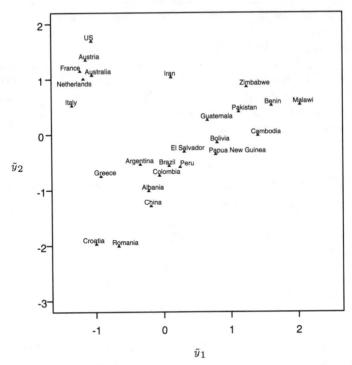

Figure 5.9 *Plot of scores on first two principal components, economic and demographic data*

The difference between CORA and PCA is that they start with different matrices. In CORA there is a symmetry: the rows and columns in the $I \times J$ matrix of Pearson residuals have the same status, whereas in the $n \times p$ matrix of standardized variables for PCA, the rows represent individuals and the columns represent variables.

5.8 Using principal component scores to replace the original variables

One use of PCA is to replace a larger set of p variables by a smaller set of q principal components. The first component, y_1, might be used alone as a univariate ($q = 1$) summary of the original variables, $x_1, \ldots, x_p$, either for use in further analyses or as an index. Indeed, the component score coefficients are sometimes used to score *new* individuals on such an index. The first two components might be used to plot the data — either scaled so that var(y_1) = λ_1, var(y_2) = λ_2, as in Figure 5.3, or with components, $\tilde{y}_1$ and $\tilde{y}_2$, standardized to unit variance as in Figures 5.8 and 5.9.

The question arises as to how much information is lost by replacing the p x's by their first q principal components; or more particularly as to how well can x_i be reconstructed from $\tilde{y}_1, \ldots, \tilde{y}_q$ for $(i = 1, \ldots, p)$.

In Section 5.6, the (standardized) principal components are given as linear functions of the (standardized) original variables,

$$\tilde{y}_j = \tilde{a}_{1j}x_1 + \cdots + \tilde{a}_{pj}x_p \qquad (j = 1, \ldots, p).$$

These equations can be inverted to give

$$x_i = a_{i1}^* \tilde{y}_1 + \cdots + a_{ip}^* \tilde{y}_p \qquad (i = 1, \ldots, p),$$

where $a_{ij}^* = \lambda_j \tilde{a}_{ij}$ is the component loading introduced in Section 5.4. Remember, this loading is the correlation between x_i and y_j. Now suppose that we try to reconstruct x_i using only the first two components. The reconstructed value is

$$\hat{x}_i = a_{i1}^* \tilde{y}_1 + a_{i2}^* \tilde{y}_2.$$

This will be close to x_i if the remaining correlations or loadings, $a_{i3}^*, \ldots, a_{ip}^*$, are all close to zero. Equivalently, we can judge how well each x_i is reproduced from the first q components by seeing how close the *communality* is to one, where the communality is the sum of the first q squared loadings so that for x_i, the communality equals

$$a_{i1}^{*2} + \cdots + a_{iq}^{*2} \qquad (i = 1, \ldots, p).$$

We shall meet the communality again in the context of factor analysis. It is the square of the multiple correlation coefficient between x_i and $y_1, \ldots, y_q$.

Table 5.12 gives the communalities for the subject mark data, for one component and for two components. History, for example, would not be adequately summarised by just the first component, but the first two components together might be judged adequate.

Table 5.12 *Communalities for one and for two components for the subject marks data*

Subject	One component	Two components
Gaelic	0.44	0.63
English	0.47	0.56
History	0.27	0.68
Arithmetic	0.54	0.72
Algebra	0.55	0.70
Geometry	0.46	0.58

5.9 Further examples and suggestions for further work

Social mobility in the UK

The correlations analyzed are taken from Ridge (1974) and are based on information provided by 713 male or female married respondents to a survey carried out in 1949 by D.V. Glass and associates at the London School of Economics. The variables relate to the respondent, her or his spouse, father, father-in law, and first born son and are described in Table 5.13. The correlations (which are all positive) are given in Table 5.14, and the loadings for the first six principal components are given in Table 5.15. You can see that many of the correlations are small and none of them are very large. Of interest is whether there are identifiable differences between the generations, and also to what extent the three measures (occupational status, further education and qualifications) are all indicators of a family's status.

Table 5.13 *Descriptions of social mobility variables*

Variable	Generation	Code	Description
x_1	1	HF/O	Husband's father's occupational status
x_2	1	WF/O	Wife's father's occupational status
x_3	2	H/FE	Husband's further education
x_4	2	H/Q	Husband's qualifications
x_5	2	H/O	Husband's occupational status
x_6	2	W/FE	Wife's further education
x_7	2	W/Q	Wife's qualifications
x_8	3	FB/FE	Firstborn's further education
x_9	3	FB/Q	Firstborn's qualifictions
x_{10}	3	FB/O	Firstborn's occupational status

Table 5.14 *Pairwise correlations ($\times$ 100) between social mobility variables*

Variable	x_1	x_2	x_3	x_4	x_5	x_6	x_7	x_8	x_9	x_{10}
x_1	100	37	23	10	43	17	13	18	8	29
x_2	37	100	23	13	38	15	10	18	10	28
x_3	23	23	100	53	35	28	28	32	25	29
x_4	10	13	53	100	24	23	38	31	35	22
x_5	43	38	35	24	100	20	14	23	11	44
x_6	17	15	28	23	20	100	47	26	12	19
x_7	13	10	28	38	14	47	100	21	19	16
x_8	18	18	32	31	23	26	21	100	50	44
x_9	8	10	25	35	11	12	19	50	100	33
x_{10}	29	28	29	22	44	19	16	44	33	100

The first six eigenvalues (explained variance) are 3.34, 1.44, 1.17, 0.89, 0.68, and 0.61, suggesting that three or four principal components should be used.

For component 1 (which explains 33% of the total variance), the loadings vary between 0.5 and 0.7, suggesting that this is a summary of the ten variables measuring the status of the family.

The second component (which explains 14% of the total variance) contrasts variables x_1, x_2, x_5, and x_{10} (which give the occupational status of different family members) and variables x_4, x_7, and x_9 (relating to qualifications) with a lesser contribution from variables x_3, x_6, and x_8 (relating to further education).

The third component (which explains a further 11% of the variance) contrasts the first born son (variables x_8, x_9, and x_{10}) with his mother (variables x_6 and x_7) and possibly other ancestors (the remaining variables).

The fourth component might also be interpretable largely as a contrast between husband and wife, but this may be pushing the data beyond its limits as the eigenvalue for this component is less than one and successive components become less reliable.

In conclusion, the correlation structure suggests that all ten variables have something in common which could be referred to as family status, but that there are differences between occupational status, further education, and qualifications, and to a lesser extent, there are differences between the generations. We shall return to this example in Chapter 6.

Table 5.15 *Loadings for first six components, social mobility data*

Variable		a_{i1}^*	a_{i2}^*	a_{i3}^*	a_{i4}^*	a_{i5}^*	a_{i6}^*
x_1	HF/O	0.50	0.56	0.15	0.07	0.06	0.61
x_2	WF/O	0.48	0.52	0.08	−0.01	0.60	−0.34
x_3	H/FE	0.68	−0.13	0.11	−0.51	−0.09	−0.06
x_4	H/Q	0.62	−0.40	0.06	−0.50	0.07	0.03
x_5	H/O	0.62	0.49	0.07	−0.13	−0.34	−0.08
x_6	W/FE	0.51	−0.24	0.52	0.44	−0.07	−0.15
x_7	W/Q	0.51	−0.41	0.50	0.21	0.06	0.11
x_8	FB/FE	0.65	−0.19	−0.44	0.27	0.03	−0.03
x_9	FB/Q	0.52	−0.35	−0.56	0.11	0.23	0.20
x_{10}	FB/O	0.65	0.22	−0.33	0.21	−0.35	−0.20
Variance		3.34	1.44	1.17	0.89	0.68	0.62
% variance explained		33.42	14.37	11.73	8.90	6.82	6.16

Educational circumstances

The correlation matrix for nine variables relating to circumstances and test results in 1964 and 1968 of girls in their fourth year of secondary school in 1968 has already been subjected to cluster analysis in Chapter 2. If you carry out a PCA of these data, you will find that there is a dominant first component, accounting for 42.6% of the total variation. When you examine the scree plot, you will notice an elbow at the second component, suggesting that only the first component is necessary. Although only the first two components have eigenvalues greater than one, components 3 and 4 have eigenvalues close to one, and by Jolliffe's criterion the first five components should be examined. Components 2, 3, and 4 have similar eigenvalues and account for 12.4%, 11.1% and 9.1% respectively of the total variance. It it therefore not clear how many components should be considered. The first three components are examined here, but you should look at the fourth to see whether this offers any further insight.

You will find that the first component is positively correlated with all variables, reflecting the mainly positive correlations in the correlation matrix (Table 2.17). The first component might therefore be interpreted as some general indicator of a girl's circumstances both at home and at school. Components 2 and 3 do not have a clear interpretation. However, some patterns emerge if we look at pairwise plots of the component loadings. Figures 5.10 and 5.11 show plots of the loadings for component 1 versus component 2, and component 1 versus component 3, respectively. From Figures 5.10 and 5.11, you can see that the loadings for variables x_1 and x_7 (parental circumstances in 1964 and 1968) are close together on all of the first three principal components. Similarly variables x_5, x_9, and x_6 (the two test scores and the type of school) have loadings close to each other on the first three principal components, but the remaining variables do not closely resemble each other or the two clusters of variables above. Compare Figures 5.10, 5.11 and Figure 2.17, which shows the results of a cluster analysis. You will find that the results of the PCA tend to confirm the cluster analysis of these data. This illustrates how the different methods of summarising a correlation matrix can reinforce each other.

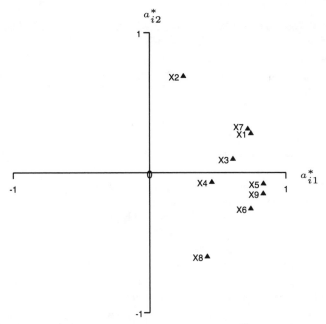

Figure 5.10 *Plot of loadings for first two components, educational circumstances*

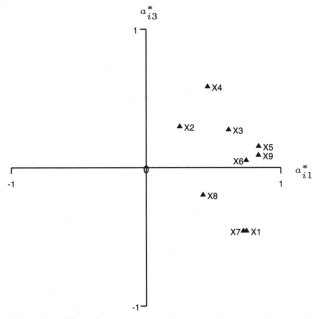

Figure 5.11 *Plot of loadings for component 1 and component 3, educational circumstances*

Television viewing in the UK

A sample of 7000 UK adults was asked whether they "really liked to watch" a range of ten television programmes (Ehrenberg 1977). The pairwise correlations between the ten variables measuring "liking to watch" the programmes are shown in Table 5.16. The programmes fall into two broad categories: sports programmes (World of Sport, Match of the Day, Grandstand, Professional Boxing, and Rugby Special) and current affairs programmes (24 Hours, Panorama, This Week, Today, and Line-Up).

The scree plot from a PCA of these data is shown in Figure 5.12. From this plot, you can see an elbow at the third component. Further, only the eigenvalues for the first two components are greater than one. You should also examine the proportions of variance explained by each component: the first component explains nearly 32% of the total variance and the first two components explain 50% of the variance. The third only accounts for an additional 9% of variance explained. All this points towards choosing the first two components to summarise the data.

The loadings for the first two components are plotted in Figure 5.13. You can see that all variables are positively correlated with the first component. The first component might therefore be interpreted as a general measure of liking to watch television. The second component has a mixture of positive and negative loadings. If you refer to the description of the programmes given above, you will find that the current affairs programmes have positive loadings on the second component while the sports programme have negative loadings. Thus this component contrasts liking to watch these two different types of programme.

Table 5.16 *Pairwise correlations between liking to watch ten television programmes*

	WoS	MoD	GrS	PrB	RgS	24H	Pan	ThW	Tod	LnU
World of Sport	1.00	.58	.62	.51	.30	.14	.19	.15	.09	.08
Match of the Day	.58	1.00	.59	.47	.33	.12	.13	.08	.04	.05
Grandstand	.62	.59	1.00	.47	.34	.14	.18	.13	.07	.08
Prof. Boxing	.51	.47	.47	1.00	.31	.12	.17	.11	.07	.09
Rugby Special	.30	.33	.34	.31	1.00	.12	.15	.06	.05	.10
24 Hours	.14	.12	.14	.12	.12	1.00	.52	.39	.24	.27
Panorama	.19	.13	.18	.17	.15	.52	1.00	.35	.20	.20
This Week	.14	.08	.13	.11	.06	.39	.35	1.00	.27	.19
Today	.09	.04	.07	.07	.05	.24	.20	.27	1.00	.15
Line-Up	.08	.05	.08	.09	.10	.27	.20	.19	.15	1.00

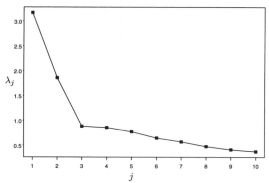

Figure 5.12 *Scree plot of eigenvalue versus number of component, television viewing data*

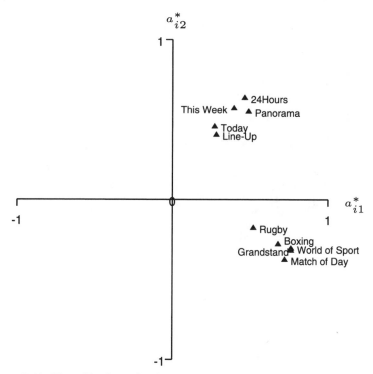

Figure 5.13 *Plot of loadings for first two components, television viewing data*

5.10 Further reading

Basilevsky, A. (1994). *Statistical Factor Analysis and Related Methods*. New York: Wiley.

Jolliffe, I. T. (1986). *Principal Components Analysis*. New York: Springer-Verlag.

CHAPTER 6

Factor Analysis

6.1 Introduction to latent variable models

Factor analysis belongs to a family of methods which involve what are called *latent variables*. Often, particularly in social science research, we cannot directly measure the variables of major interest. Examples of such concepts are intelligence, political attitude (left-wing, moderate, or right-wing), and socio-economic status. Although we use these concepts in social science discourse as if they were just like any other variable, they differ from other variables in that they cannot be observed — which is why they are called latent. In some cases, a concept may be represented by a single latent variable, but often they are multidimensional in nature and so involve more than one latent variable. Suppose that there are q latent variables, denoted by $y_1, y_2, \ldots, y_q$. These latent variables are commonly called *factors* (many writers use f rather than y to denote factors). Latent variable methods, of which factor analysis is the oldest and most widely used, form the subject of this and the next three chapters.

There is a close link between factor analysis and principal components analysis. In fact, it is common to regard PCA as a method of factor analysis. Some books (for example, Basilevsky 1994 and the SPSS computer package) treat both methods within the same framework. We shall explain the justification for this at the end of the chapter but we prefer to keep them quite distinct at this stage and for two reasons. We have introduced PCA as a descriptive method concerned with summarising a data matrix in a manner which expresses its structure in a small number of dimensions. Factor analysis, on the other hand, is a model-based technique. That is, it involves assumptions about the joint distributions over some relevant population of the variables involved. This enables us to make inferences about the population using the notions of goodness-of-fit, statistical significance and the precision of estimation. We link the observable to the unobservable variables by a probability model as we shall see later.

This point in the book marks the transition from the descriptive methods of the earlier chapters to the model-based methods which follow.

The second reason for emphasizing the difference between PCA and factor analysis is that we want to emphasize the strong link between factor analysis and the other latent variable methods treated in the following three chapters. Traditionally, latent trait (which is factor analysis for categorical data) and latent class analysis have been treated quite separately from factor analysis. Their essential unity has been obscured by the use of different notation and the practices of the rather different scientific cultures in which they have been used. They differ in the level of measurement used for the variables involved but share a common basis of interpretation which we shall seek to emphasize.

In order to measure the latent variable of interest, we often collect observable variables which we feel are likely to be indicators of the latent variable(s). Suppose that we collect p observed variables, denoted by $x_1, x_2, \ldots, x_p$. The xs are also called *indicators, items* or *manifest variables* .

Examples of problems involving latent and manifest variables

i) There is great interest in measuring intelligence. This is conceived to be an important characteristic of individuals which they possess to a greater or lesser extent. However, it is not like weight or age for which there is some ready-made measuring instrument. Intelligence is a construct; that is, it is a concept which we find useful and meaningful but which does not exist in the sense that more tangible properties such as weight do. We can, however, introduce it into a mathematical model and treat it just like any other variable. Intelligence is a good example of a latent variable. The indicator variables in this case are quantities which are presumed to be influenced by the latent variable. In this case, these are usually the scores obtained in a series of test items chosen because it is believed that more intelligent people will perform better. Some items might be verbal or arithmetical, others might involve spatial exercises designed to test the ability to see patterns. If the items all require the same sort of basic mental ability, we would expect the scores on the items to be positively correlated. The problem is to see whether that correlation can be accounted for by a single latent variable and, if so, how we can determine where to place individuals on the latent scale.

ii) The measurement of political attitude is very similar to the case of intelligence. We describe individuals as left- or right-wing and some, for example, as more right-wing than others. Implicit in this kind of language is the idea that there is a scale on which individuals can be located extending from extreme left at one end to extreme right at the other. This is a latent scale and if we are to construct such a scale, we require appropriate indicators. These could be provided by a social survey in which respondents are asked their attitudes on a range of political issues, for example, private healthcare, private education and trade unions.

iii) In order to measure a latent variable such as the socio-economic status of a household, we might similarly collect information about household income, and the occupations and education levels of household members.

In each of these examples, we have used our intuitive understanding of the latent variable in which we are interested to identify some manifest variables which, we expect, will reveal something about that underlying latent variable. In effect, we have started with a latent variable and looked for manifest variables which would serve as indicators because we already had an idea of what the key latent variables would be. Sometimes, we proceed in the opposite direction. If, for example, we have a large general purpose survey we might wonder whether the large number of manifest dimensions represented by, perhaps, 50 questions could be reduced to a small number of dimensions without significant loss of information. This second approach is essentially the one we follow when using PCA. In practice, the true situation usually lies somewhere between these two extremes. The enquiry may have been motivated by the desire to investigate the existence of some latent variables, but we wish to carry out the investigation in an open enough fashion for unexpected features to be picked up.

Latent variable models

Latent variable models are closely related to the standard regression model. It may, therefore, be helpful to describe the central idea of factor analysis in terms of regression analysis. A regression model expresses the relationship between a dependent variable and one or more independent, or regressor, variables. In factor analysis, the regression relationship is between a manifest variable and the latent variables. In both cases, we add distributional assumptions about the residual or error terms which enable us to make inferences. The essence of the problem that factor analysis, or other latent variable analysis, has to solve is that of inverting the regression relationships to tell us about the latent variables when the manifest variables are given. Since we can never observe the latent variables, we can only ever learn about this relationship indirectly.

Several manifest variables will usually depend on the same latent variable, and this dependence will induce a correlation between them. Indeed, the existence of a correlation between two indicators may be taken as evidence of a common source of influence. As long as any correlation remains, we may therefore suspect the existence of a further common source of influence. The aim of a latent variable analysis is to determine whether the dependencies between the observed variables may be explained by a small number of latent variables. As we observed above, latent variable models may be used either in an exploratory way to identify the latent variables underlying a set of items, or in a confirmatory way, to test whether a set of items designed to measure particular concepts does indeed reveal the assumed structure.

There are various types of latent variable model. These models are distinguished by the level of measurement of the observable variables and the assumptions made about the level of measurement of the latent variables. Table 6.1 shows a classification of latent variable models.

This table does not exhaust the possibilities because, for example, the manifest variables may be a mixture of metrical and categorical variables. However, this classification is sufficient for the purposes of this book.

We begin, in this chapter, with a discussion of *factor analysis*, which is an appropriate technique when all observed variables are measured on a metrical

Table 6.1 *Classification of latent variable models*

| Latent variables (y) | Observed variables (x) | |
	Metrical (interval/ratio)	Categorical (nominal/ordinal)
Metrical (interval/ratio)	Factor analysis	Latent trait analysis
Categorical (nominal/ordinal)	Latent profile analysis	Latent class analysis

(interval or ratio) scale. The factor model assumes that the latent variables are also metrical.

6.2 The linear single-factor model

The simplest factor model is one which involves only one factor. Charles Spearman, who invented factor analysis (Spearman, 1904), introduced this model in the study of human intelligence. For rather special reasons connected with that particular application, he referred to it as a two-factor model, but that usage has long been abandoned.

We introduce the model by means of a practical example which thus forms a bridge from what is familiar to what is novel. This will serve to show that a factor analysis model is simply a set of regression models in which some of the variables (the latent variables) are unobserved. By repeating the argument of the last section in relation to a special case, the central ideas should be made clearer.

Factor analysis aims to explain the correlations among the set of manifest variables. Such correlations are often spurious in the sense that there is no direct causal link between the variables concerned. They sometimes arise because the variables in question have a common dependence on one or more other variables. For example, the fact that the size of a child's feet are positively correlated with his or her writing ability does not mean that large feet help the child to write better. The correlation is, rather, an incidental consequence of the fact that both are correlated with age — the older the child, the larger the feet and the better the handwriting. When one finds correlations among variables like this, it is important to investigate whether they can be explained by a common dependence on some other variables.

In some circumstances, there may be an obvious candidate for the role of an "other variable". Suppose, for example, that we look at weekly family expenditure for a large sample of families on a variety of things: food, travel, entertainment, clothes, and so on. Suppose also that we find that the correlations (between pairs of purchases) are positive. It would not be credible to claim that high spending on clothes, say, *causes* high spending on travel. It seems more plausible to suppose that high spending on any of these things is a consequence of having a high income. To investigate this hypothesis, we might obtain further data on the incomes of each family. This would enable us to see whether the size of each item of expenditure was related to total income and, if so, whether that relationship wholly explained the correlations between expenditures.

How might we investigate this empirically? One way would be to specify how each expenditure might be related to income. To get a preliminary idea of how to do this, we might plot expenditure on food against income. Suppose it turned out to be roughly linear and that a similar result was obtained for each other item. We could then fit simple regressions of the form:

$$C_i = \alpha_i + \beta_i I + e_i \qquad (i = 1, 2, \ldots), \qquad (6.1)$$

where C_i is consumption or expenditure on the ith item, I is the income of the family, α_i and β_i the intercept and slope, respectively, of the regression, and e_i a random component or residual, specific to C_i with mean zero, independent of I, which explains the residual variation about the line. If we found that this model was a good fit for all items of expenditure, and that the residual e_is were uncorrelated with each other, then we would have shown that income was the only detectable determinant of expenditure. For fixed income, expenditure on item i would behave like a random quantity with mean $\alpha_i + \beta_i I$ and standard deviation given by the standard deviation of e_i and, because the residuals are independent, all correlation between the observed variables would have been removed. If all this proved to be the case (and there are many ifs) we would be satisfied that the mutual correlations among the original expenditures were explained by their common dependence on income. Furthermore, the regression coefficients, β_i, would tell us how strongly each item of expenditure depended on income.

In most practical problems, there is no ready-made variable, like the income of this example, to invoke as an explanation. (Even if there were, it might be impractical to collect it because, for example, the question was held to be too intrusive.) In the absence of any such observable variable, we have to ask whether there could be *any* such latent variable (or variables) which could play the same role.

Whether or not the latent variable is a real variable, which we are unable to observe, or a construct, we are faced with the same fundamental question: is there any way of estimating the regression models of (6.1) without knowing the values of I? This is the technical problem which factor analysis seeks to solve. We shall see below that, rather surprisingly, the set of correlations does contain enough information to enable us to estimate the regression relationships and hence to infer that there could be some common factor.

Suppose that our p manifest variables, $x_1, x_2, \ldots, x_p$ are believed to depend on a single factor or latent variable, y. The simplest way to express the regression of each x on y is by means of the linear model,

$$x_i = \alpha_i + \beta_i y + e_i \qquad (i = 1, 2, \ldots, p), \qquad (6.2)$$

y may be called the *common factor* since it is common to all x_is. The e_is were sometimes called *specific* or *unique factors*, since they are unique to a particular x_i. (It was because Spearman thought of the e_i as factors that he called his model a two-factor model. Modern terminology counts only the number of common factors.) In the one-factor model, we make the usual regression assumption that e_i is independent of y and has a normal distribution with mean zero and standard deviation σ_i. We also assume that $e_1, e_2, \ldots, e_p$ are independent so that $x_1, x_2, \ldots, x_p$ are conditionally independent, given y. Then we can make some deductions about the distribution of the xs and, in particular, about their covariances and correlations. We can choose the scale and origin of y as we please because this does not affect the form of the regression equation so we choose to make y have zero mean and unit standard deviation. It turns out that, under this model, the theoretical covariance coefficients have a very simple form, namely

$$\text{Cov}(x_i, x_t) = \beta_i \beta_t \qquad (i, t = 1, \ldots, p; \; i \neq t).$$

The important thing about this formula is that the covariance is a product of two numbers, one depending only on i and the other only on t. From these equations, it is possible to deduce something about the regression coefficients in the model. For example,

$$\text{Cov}(x_1, x_2)\text{Cov}(x_2, x_3)/\text{Cov}(x_1, x_3) = \beta_2^2,$$

which serves to determine β_2 from the covariances. However, we can construct other expressions like this which should also give β_2: for example, if we replace the subscripts 1 and 3 by any other pair in the range 1 to p, the right hand side will be the same. If the model is correct and if we knew the true covariances, $\text{Cov}(x_i, x_t)$, then the different equations would all give exactly the same value of the regression coefficient, β_2.

Since, in analysing real data, we would only have estimated or fitted covariances (denoted by $\text{cov}(x_i, x_t)$ with a lower case "c"), we would not get identical estimated values $\hat{\beta}_i$ of β_i even if the model were correct. (Throughout this chapter and the following chapters, we will use a "hat" above a parameter to denote its estimated value.) However if all the "estimates" of β_i were similar, that would suggest that the model was a good fit. In the early days, factor models were fitted by a method very similar to this, tedious to apply and not easily extended to the case of several factors, but exploiting the basic result behind fitting all factor models, which is that we can determine the parameters of the model from the covariances between the manifest variables without knowing the values of the factors themselves.

The one-factor model can easily be extended to allow an arbitrary number of factors. We simply replace the simple linear regression equation with a multiple regression equation. In doing so, we shall introduce a more flexible notation and terminology which will also be useful for the models of the next three chapters.

6.3 The general linear factor model

The general linear factor model for p observed variables and q factors or latent variables takes the form:

$$x_i = \alpha_{i0} + \alpha_{i1}y_1 + \alpha_{i2}y_2 + \cdots + \alpha_{iq}y_q + e_i \qquad (i = 1, \ldots, p), \qquad (6.3)$$

where $y_1, y_2, \ldots, y_q$ are the common factors or latent variables, e_i are residuals, and α_{i1}, α_{i2}, and α_{iq} are called the *factor loadings*. The constant term α_{i0} plays no role in fitting or interpreting the model; it can be dispensed with if we assume the xs are measured about their mean. The other αs play a key role in interpreting the factors. For this purpose, it is useful to know that the factor loadings turn out to be the covariances between the latent variables and the xs (or correlations if the xs are standardized). As in the simple model, we scale and locate the ys so that they have mean zero and standard deviation one.

The linear factor model has been based on the idea of multiple linear regression, but it is more complicated in that instead of having just one response or criterion variable it has p which, conditionally, are mutually uncorrelated given the explanatory variables and, furthermore, the explanatory variables are unobserved.

We list the assumptions of the model as follows:

i) $y_1, y_2, \ldots, y_q$ are uncorrelated with each other (though we relax this assumption later — see Section 6.6),

ii) $y_1, y_2, \ldots, y_q$ each have a mean of zero and variance of one,

iii) $e_1, e_2, \ldots, e_p$ are uncorrelated with each other,

iv) each e_i has a mean of zero, but they may have different variances, $\mathrm{Var}(e_i) = \sigma_i^2$, $(i = 1, \ldots, p)$,

v) the ys are uncorrelated with the es.

Sometimes, and for some purposes, we make the following additional assumptions:

vi) $y_1, y_2, \ldots, y_q$ follow a multivariate normal distribution,

vii) $e_1, e_2, \ldots, e_p$ follow a multivariate normal distribution.

Assumptions vi) and vii) imply that $x_1, x_2, \ldots, x_p$ also have a multivariate normal distribution. These assumptions lead to the *normal linear factor model*. Assumptions (iii) and (v) imply that the correlations among the xs are wholly explained by the factors.

Properties of the linear factor model

An alternative notation for writing the general linear factor model, given by equation (6.3) and assumptions (i) through (v) above is:

$$
\begin{aligned}
\mathrm{E}(x_i \mid \mathbf{y}) &= \alpha_{i0} + \alpha_{i1}y_1 + \cdots + \alpha_{iq}y_q & (i = 1, 2, \ldots, p), \\
\mathrm{SD}(x_i \mid \mathbf{y}) &= \sigma_i & (i = 1, 2, \ldots, p), \\
\mathrm{Cov}(x_i, x_t \mid \mathbf{y}) &= 0 & (i, t = 1, 2, \ldots, p; \ i \neq t).
\end{aligned}
$$

where $E(x_i \mid \mathbf{y})$ is read as the conditional expectation (or mean value) of x_i for fixed $\mathbf{y}$ (i.e. for fixed values of $y_1, y_2, \ldots, y_q$). Similarly, $SD(x_i \mid \mathbf{y})$ is the conditional standard deviation of x_i given $\mathbf{y}$, which is, of course, just the standard deviation of e_i. The final statement states that the conditional covariance is zero. This implies that the conditional correlation is zero.

Consider the earlier example of children's writing: if x_1 is foot size, x_2 is writing ability and y is the single variable, age, then x_1 and x_2 are positively correlated, but *conditional* on y they are uncorrelated:

$$\mathrm{Corr}(x_1, x_2) \;>\; 0,$$
$$\mathrm{Corr}(x_1, x_2 \mid y) \;=\; 0.$$

Differences in age fully account for the apparent correlation between foot size and writing ability.

If the conditional distributions of $x_1, \cdots, x_p$ given $\mathbf{y}$ are normal, then zero conditional correlation implies conditional independence. The normal linear factor model is a conditional or local independence model.

Returning to the general linear factor model, we deduce that the unconditional or marginal mean of x_i is:

$$E(x_i) = \alpha_{i0} \qquad (i = 1, 2, \ldots, p), \tag{6.4}$$

that the unconditional or marginal variance is:

$$\mathrm{Var}(x_i) = \alpha_{i1}^2 + \alpha_{i2}^2 + \cdots + \alpha_{iq}^2 + \sigma_i^2 \qquad (i = 1, 2, \ldots, p), \tag{6.5}$$

and that the unconditional covariance between two observed variables, x_i and x_k, takes the form:

$$\mathrm{Cov}(x_i, x_k) = \alpha_{i1}\alpha_{k1} + \alpha_{i2}\alpha_{k2} + \cdots + \alpha_{iq}\alpha_{kq}, \tag{6.6}$$

where $(i, k = 1, 2, \ldots, p;\ i \neq k)$.

Therefore, the variance is composed of two parts: $\alpha_{i1}^2 + \alpha_{i2}^2 + \cdots + \alpha_{iq}^2$, the part of the variance of x_i explained by the common factors (also called the *communality*), and σ_i^2, the residual or specific variance. The covariances between the xs depend solely on the regression coefficients linking them with the common factors. If the common factors are held constant there will be no remaining source of covariance among the xs.

From the above expressions for the variances and covariances of the xs, we obtain the form of the covariance matrix assumed under the factor model. For example, under the one-factor model the covariance matrix of the xs is:

$$\begin{pmatrix} \alpha_{11}^2 + \sigma_1^2 & \alpha_{11}\alpha_{21} & \cdots & \alpha_{11}\alpha_{p1} \\ \alpha_{21}\alpha_{11} & \alpha_{21}^2 + \sigma_2^2 & \cdots & \alpha_{21}\alpha_{p1} \\ \vdots & & & \vdots \\ \alpha_{p1}\alpha_{11} & \alpha_{p1}\alpha_{21} & \cdots & \alpha_{p1}^2 + \sigma_p^2 \end{pmatrix}.$$

In the general case, the elements in this matrix are replaced by the expressions given in (6.5) and (6.6) above.

Whereas in PCA, the choice of the scale of the variables changes the components — variables with large variances tending to dominate the first few components — in FA, because it is model-based, the factors are the same whatever the scale of measurement used for the observable variables. However, it is common practice to scale or standardize the variables to zero mean and unit variance. As we have already chosen to standardize the latent variables to unit variance, this will give factor loadings on the scale of correlations. The factor loading between observable variable and a factor will be their correlation. Hence the communality between an observable variable and the set of factors will also be the squared multiple correlation coefficient, R^2. Thus, using standardized variables (and therefore analysing the correlation matrix rather than the covariance matrix) makes the interpretation of the results easier.

Fitting the model

The usual starting point for a factor analysis is the correlation matrix of the xs. The correlations should first be examined. If the correlations between the xs are low then factor analysis is unlikely to be useful since the xs are unlikely to share common factors. Inspection may also reveal interesting patterns or undesirable anomalies of various kinds. For example, if two very highly correlated xs have been included inadvertently, meaning that one adds very little information to the other, the correlation close to one will be immediately apparent. The problem here is that factors common to the other xs would not explain this particularly high correlation and we would not wish to fit an extra factor just to explain one correlation.

Fitting the factor model itself involves finding the values of the parameters which make the observed correlation matrix as close as possible to that predicted by the model. In the case of the one-factor model, we saw (Section 6.2) that this could be done by a rather *ad hoc* procedure. What we need is a numerical routine which can be programmed to fit any model. Inspection of any text or computer software package will reveal a bewildering array of methods with names like ordinary least squares, generalised least squares and maximum likelihood. All of these methods start by constructing a measure of the distance between the observed and predicted correlation matrices; they differ in the measure they choose. Ordinary least squares, as its name suggests, simply sums the squares of the differences between the corresponding elements of the two matrices. Maximum likelihood uses a distance which arises naturally when we make the normality assumptions (vi) and (vii) in Section 6.3, but it can still be used when this is not the case. In practice, one usually finds that all methods give rather similar results, and it is instructive to try several methods since all are very fast on desk top computers. There is some theoretical advantage in using either maximum likelihood or weighted least squares.

Fitting the model does not, of course, guarantee that the fit will be acceptable. We shall describe methods of judging the suitability of a model below.

6.4 Interpretation

The factor loadings

The factor loadings have a similar interpretation to the component loadings in PCA. If the correlation matrix is analyzed, and if the factors are constrained to be uncorrelated (assumption (i), Section 6.3), the factor loading $\hat{\alpha}_{ij}$ is the correlation between the observed variable x_i and the latent variable y_j. A factor may be interpreted, or labelled, by examining the pattern of the loadings on that factor across the observed variables. To illustrate the interpretation of factor loadings, we re-analyze two datasets that were previously analyzed using PCA.

The linear factor model with two factors was fitted to the subject marks data described in Section 5.5. The estimated loadings obtained by the maximum likelihood method are shown in Table 6.2. Since the correlation matrix of marks was analyzed, the loadings may be interpreted as correlations between the mark in a subject and a factor. For example, the correlation between Gaelic and the first factor is estimated as 0.56. In attempting to interpret the factor, we have to ask ourselves what it is that is correlated positively and fairly strongly with each of the subject marks. The position is very similar to the one we faced when interpreting the component loadings obtained from PCA for this set of data. Therefore, we may interpret the first two factors in the same way as the first two components. The first factor measures overall ability in the six subjects, while the second contrasts humanities and mathematics subjects.

Table 6.2 *Estimated factor loadings from a two-factor model of the subject marks data*

Subject	$\hat{\alpha}_{i1}$	$\hat{\alpha}_{i2}$
Gaelic	0.56	0.43
English	0.57	0.29
History	0.39	0.45
Arithmetic	0.74	−0.28
Algebra	0.72	−0.21
Geometry	0.60	−0.13

A two-factor model was also fitted to the children's personality trait data described in Section 5.5. The loadings are shown in Table 6.3. Again, the interpretation of the factors is essentially the same as that of the principal components. The first factor represents some overall personality measure, while the second contrasts indicators, such as sociability, of how a child relates to other people with those which are internal to the individual, like guilt.

Table 6.3 *Estimated factor loadings from a two-factor model of the children's personality trait data*

Variable (personality trait)	$\hat{\alpha}_{i1}$	$\hat{\alpha}_{i2}$
Mannerliness	0.65	0.57
Approval seeking	0.54	0.54
Initiative	0.61	−0.45
Guilt	0.63	−0.54
Sociability	0.56	0.54
Creativity	0.72	−0.59
Adult role	0.67	−0.45
Cooperativeness	0.64	0.60

Communalities

The communality of a standardized observable variable is the squared multiple correlation coefficient or the proportion of the variance that is explained by the common factors. The estimated communalities from the factor analysis of the subject marks data are shown in Table 6.4. These show, for example, that 49% of the variance in Gaelic scores is explained by the two common factors. Recall also from Section 6.3 that the communality of a variable is calculated as the sum of the squared loadings for that variable. For example, the communality for Gaelic scores is calculated as $0.56^2 + 0.43^2 = 0.49$. The larger the communality, the better does the variable serve as an indicator of the associated factors. Or, put another way, a variable, x_i, with a large communality is a "purer" indicator of the common factors, y, with less contamination from the specific component, e_i. The sum of the communalities is the variance explained by the factor model. From Table 6.4, this is 2.81 or 47% of 6 which is the total variance for the subject marks data.

Table 6.4 *Communalities from fitting a linear two-factor model to the subject marks data*

	Communalities
Gaelic	0.49
English	0.41
History	0.36
Arithmetic	0.62
Algebra	0.56
Geometry	0.37

6.5 Adequacy of the model and choice of the number of factors

A primary goal of FA is to reduce the dimensionality of the multivariate data set while retaining sufficient dimensions to provide a good approximate representation of the original data. There are several ways in which the adequacy of a factor model may be assessed.

i) *Percentage of variance explained by the factors*

Although the aim of factor analysis is to explain the covariances or equivalently the correlations between the observed variables rather than their variances, the proportion of variance explained by the common factors should be reasonably high. The two common factors fitted to the subject marks data together explain approximately 47% of the total variance, which is roughly the same as the first principal component. Also the communalities can be used to check that the individual observable variables are adequately explained by the factors. From Table 6.4, it appears that the Arithmetic marks are better explained than the History marks.

ii) *Reproduced correlation matrix*

A good way of assessing the fit of a model is to compare the fitted (reproduced) correlation matrix of the xs with the correlation matrix computed from the sample data. Table 6.5 shows the reproduced correlation matrix obtained from fitting a two-factor model to the subject marks data. The diagonal entries of the upper section of the table are the communalities (also given in Table 6.4). The off-diagonal entries of this section of the table are the reproduced correlations. For example, the correlation between Gaelic and English marks is estimated from the model as

$$\text{corr}(x_2, x_1) = \hat{\alpha}_{21}\hat{\alpha}_{11} + \hat{\alpha}_{22}\hat{\alpha}_{12} = (0.57 \times 0.56) + (0.29 \times 0.43) = 0.44.$$

The reproduced correlations should be compared with the sample correlation matrix given in Table 5.2. The lower section of the table shows the discrepancies or differences between the observed sample correlations and the reproduced correlations. Here, the differences are small suggesting that the two-factor model is a good fit.

iii) *Goodness-of-fit test*

If the normal factor analysis model is assumed, we can carry out a log-likelihood-ratio test or goodness-of-fit test to test the null hypothesis that the covariance matrix of the xs has the form specified by the factor model. Failure to reject this null hypothesis would imply a good fit. The test statistic, denoted by W, has a chi-squared distribution under the null hypothesis with $\{(p-q)^2 - (p+q)\}/2$ degrees of freedom.

The test statistic for the two-factor model fitted to the subject marks data was 2.18 on 4 degrees of freedom, suggesting that the model is a very good fit.

If a model with a given number of factors is deemed to be a poor fit, more factors may be added until a good fit is achieved. However, as always, we have to bear in mind the balance between interpretability and goodness-of-fit. A well-fitting model with a large number of factors may not be interpretable, while a poorer-fitting model may still reveal some interesting features of the data. For large sample sizes, the test becomes sensitive to small departures from the model which may not be of practical relevance.

Also, a statistically significant result may be due to departures from multivariate normality rather than from the need for an extra factor.

iv) *Standard errors of factor loadings*

Traditionally, standard errors of factor loadings have not been routinely quoted and some packages still do not give them. However, both in interpreting the factors and in deciding how many factors are needed, it would be useful to examine the standard errors. For example, if the absolute values of the estimated loadings for a factor were all less than twice as large as their standard errors, then the imprecision of that factor would render it useless. As with the goodness-of-fit test statistic, it is necessary to know the sample size in order to calculate standard errors. There is a theoretical aspect that needs to be taken into account when standard errors are computed. Software for structural equation models (LISREL, EQS, M-Plus) provides standard errors for the factor loadings of the one-factor model but not for the model with more than one factor unless a confirmatory factor analysis model is fitted. The reason is that no unique solution exists when the number of factors is greater than one (see Section 6.6). A unique solution can be obtained by fixing some of the factor loadings to a pre-specified value. The number of factor loadings to be fixed depends on the number of factors fitted. For example, in the two-factor model one loading needs to be fixed to obtain a unique solution.

Choosing the number of factors

The number of factors, q, must be small enough for the degrees of freedom $[(p-q)^2 - (p+q)/2]$ to be greater than or equal to zero. So when $p = 3$ or $p = 4$, q cannot be greater than one, but when $p = 20$, q could be as large as 14. In choosing how many factors to fit, a useful first step is to carry out a principal components analysis because the number of components needed is often a good guide to the number of factors. The number of principal components required is judged according to the criteria described in Section 5.4. A factor model with the same number of factors can then be fitted. The rationale for this procedure is given in Section 6.10 where we investigate the relationship between PCA and factor analysis in more detail. To assess the adequacy of a model with a given number of factors, we use the percentage of variance explained, the communalities, the discrepancies between the observed and reproduced correlations, the goodness-of-fit test and the standard errors of the estimated factor loadings, as described above.

Table 6.5 *Reproduced correlations and communalities (top section) for a linear two-factor model fitted to the subject marks data, and discrepancies between observed and reproduced correlations (bottom section), subject marks data*

Correlation	Gaelic	English	History	Arithmetic	Algebra	Geometry
Gaelic	0.49	0.44	0.41	0.29	0.31	0.28
English	0.44	0.41	0.35	0.34	0.35	0.30
History	0.41	0.35	0.36	0.16	0.19	0.17
Arithmetic	0.29	0.34	0.16	0.62	0.59	0.48
Algebra	0.31	0.35	0.19	0.59	0.56	0.46
Geometry	0.28	0.30	0.17	0.48	0.46	0.37
Discrepancy						
Gaelic		0.00	0.00	0.00	0.02	−0.03
English	0.00		0.00	0.01	−0.03	0.03
History	0.00	0.00		0.00	0.00	0.00
Arithmetic	0.00	0.01	0.00		0.00	0.00
Algebra	0.00	−0.03	0.00	0.00		0.00
Geometry	−0.03	0.03	0.00	0.00	0.00	

There are also other formal methods of choosing the optimum number of factors based on what are called information or model selection criteria (Akaike, Bayesian etc.). A discussion of those criteria can be found in Sclove (1987).

6.6 Rotation

When we posed the problem of fitting a factor model, we tacitly assumed that there was just one set of parameter values which would minimise the chosen measure of distance between the observed and predicted correlation matrices. This is true for the one-factor model, but with two or more factors there are infinitely many sets of values which all give the same minimum distance. At first sight this seems to be a serious drawback, but it allows us to introduce other criteria for choosing among solutions. However, this must not be seen as granting a license to pick and choose among the solutions until one finds one which suits one's preconceptions. A criticism often levelled against factor analysts is that subjectivity seems to play too big a role. This is a misreading of the situation in two respects. First, one cannot obtain *any* solution one wants. Secondly, the situation is more accurately described as one in which the *same* solution can be expressed in different ways. In fact, certain features, such as the communalities, remain the same in all versions of the solution.

Rotation is the name given to the process by which we move from one solution to another. The name comes from the geometrical representation of the procedure.

Once a factor model is fitted, the factors may be transformed to give a new set of factors, say $y_1^*, y_2^*, \ldots, y_q^*$. In the process, the estimated factor loadings are also transformed to give $\hat{\alpha}_{i1}^*, \hat{\alpha}_{i2}^*, \ldots, \hat{\alpha}_{iq}^*$. Our model specifies (assumption (i), Section 6.3) that $y_1, \ldots, y_q$ are mutually uncorrelated. Following an *orthogonal* transformation, the transformed $y_1^*, \ldots, y_q^*$ will also be mutually uncorrelated. Geometrically, the axes will have been rotated while being kept at right angles.

However, sometimes an oblique rotation might yield transformed factors that are easier to interpret. In such a case, assumption (i) in Section 6.3 would need to be relaxed to allow the transformed factors to be correlated.

Factor rotation is used to clarify the underlying structure of the factors. The usual motivation is to find a pattern of loadings which is relatively easy to interpret. One of the most useful patterns is described as *simple structure*. The loadings are said to have simple structure if each variable has a large contribution from only one factor, with close to zero contributions from the other factors. An illustration of simple structure for a three-factor model fitted to eight observed variables is shown in Table 6.6. The observed variables have been partitioned into three groups, each associated with one of the latent variables. For example, the third factor has large positive loadings on x_4, x_5 and x_6. This factor may be interpreted by thinking about what these three variables have in common exactly as if we had done the analysis on these variables alone. In effect, we are reducing the interpretation problem to the one we faced with a single factor.

Table 6.6 *Illustration of factor loadings with simple structure from a three-factor model*

	α_{i1}	α_{i2}	α_{i3}
x_1	+	-	-
x_2	-	+	-
x_3	+	-	-
x_4	-	-	+
x_5	-	-	+
x_6	-	-	+
x_7	+	-	-
x_8	-	+	-

+ indicates a large positive loading
- indicates a small, close to zero loading

There is no guarantee, of course, that it will be possible to find a solution with something close to simple structure. But if we can find one, it will make the interpretation that much easier. The rotation routines provided by the various software packages are designed to search for that solution among the solution set which is as close as possible to simple structure.

Illustration of orthogonal rotation in a two-factor model

The unrotated factor loadings from fitting a two-factor model to a fictional set of seven observed variables are given in the first two columns of Table 6.7. These loadings are plotted in Figure 6.1. An orthogonal rotation is applied to the axes in Figure 6.1 with the aim of achieving a new set of loadings with simple structure. In geometrical terms, this means that we are looking to rotate the axes so that the points lie close to one or other of the axes. The dashed lines in Figure 6.1 represent rotated axes having this property. In this case, almost perfect simple structure has been achieved through rotation. The coordinates of the points with respect to these rotated axes are the new factor loadings given in the last two columns of Table 6.7. The rotated factor y_1^* contributes to variables x_5, x_6 and x_7 but makes virtually no contribution to the first four variables. The rotated factor y_2^* contributes to the first four variables, but makes almost no contribution to x_5, x_6, and x_7.

Table 6.7 *Unrotated and rotated factor loadings from a two-factor model*

	Unrotated		Rotated	
	$\hat{\alpha}_{i1}$	$\hat{\alpha}_{i2}$	$\hat{\alpha}_{i1}^*$	$\hat{\alpha}_{i2}^*$
x_1	0.2	0.3	0.0	0.4
x_2	0.4	0.5	0.0	0.6
x_3	0.6	0.7	0.0	0.9
x_4	0.7	0.7	0.0	1.0
x_5	0.5	−0.5	0.7	0.0
x_6	0.7	−0.6	0.9	0.0
x_7	0.3	−0.2	0.4	0.0

It is important to note that rotation does not alter the fit of the model. Rotation does not change either the reproduced correlation matrix or the goodness-of-fit test statistic. The communalities also remain unchanged. This is because rotation has not changed the relative positions of the loadings. In the plot of the loadings, loadings that appear close together before rotation also appear close together after rotation. However, since rotation alters the loadings, the interpretation of the new factors will be different. Also, although the overall percentage of variance explained by the common factors remains the same after rotation, the percentage of variance explained by each factor will change. Rotation redistributes the explained variance across the factors.

Procedures for orthogonal rotation

In the example above, it is possible to examine a plot of the unrotated loadings in order to find a suitable rotation that will lead to simple structure. However, it is not always clear from the loadings plot which rotation should be carried out, particularly if there are more than two factors in the model.

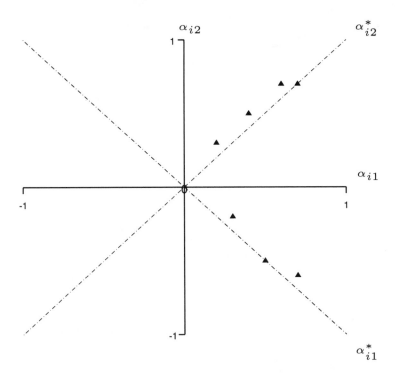

Figure 6.1 *Plot of rotated factor loadings from a two-factor model*

Some procedures have been developed to search automatically for a suitable rotation. For example, the VARIMAX procedure attempts to find an orthogonal rotation that is close to simple structure by finding factors with few large loadings and as many near-zero loadings as possible.

Non-orthogonal (oblique) rotations

Sometimes simple structure can by achieved only be means of a non-orthogonal (oblique) rotation. This type of rotation requires us to relax the original assumption of the linear factor model that the latent variables be uncorrelated. An oblique rotation leads to correlated factors. Although this complicates the interpretation of the factors, it is often reasonable to expect the latent variables to be correlated. For example, one might expect a child's mathematical ability to be positively correlated with their verbal ability. In that case, a factor analysis that assumes the latent variables to be uncorrelated may not uncover latent variables representing mathematical ability and verbal ability. Figure 6.2 shows how an oblique rotation for the subject marks data can produce new oblique axes, one going through the cluster of History, Gaelic, and English, and the other through the cluster Geometry, Algebra, and Arithmetic. The correlation between these transformed factors is 0.515.

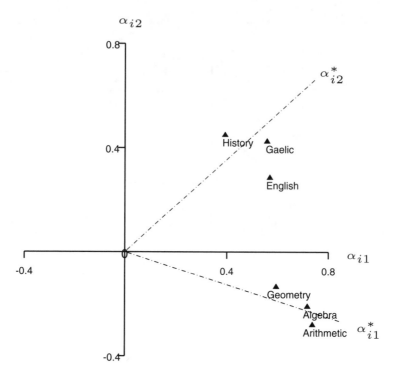

Figure 6.2 *Plot of unrotated and rotated factor loadings for the subject marks data*

6.7 Factor scores

Sometimes, we want to calculate an individual's score on the latent variable(s), perhaps for use in subsequent analysis. Such scores can be determined exactly in PCA by the simple expedient of substituting the values of an individual's manifest variables into the expressions for the principal components (see Section 5.6). In factor analysis it is not so straightforward, because the factors are random variables which have a probability distribution. There are various methods for obtaining predicted factor scores. All of them lead to expressions of the form:

$$
\begin{aligned}
y_1 &= c_{11}x_1 + c_{12}x_2 + \cdots + c_{1p}x_p \\
y_2 &= c_{21}x_1 + c_{22}x_2 + \cdots + c_{2p}x_p \\
&\vdots \\
y_q &= c_{q1}x_1 + c_{q2}x_2 + \cdots + c_{qp}x_p .
\end{aligned}
$$

The cs are called estimated factor score coefficients (the different methods giving rise to different cs). We shall use the Thomson or regression estimates which are estimates of the conditional expectations or mean values of the factor given the values of the manifest variables. Estimated factor score coefficients for computing scores for the two factors for the subject marks data are given in Table 6.8. To obtain the score for an individual boy, we would need to know his marks, $x_1, \ldots, x_6$.

Table 6.8 *Coefficients for calculating factor scores (regression method) for the subject marks data*

Subject	c_{1i}	c_{2i}
Gaelic	0.20	0.39
English	0.17	0.23
History	0.11	0.33
Arithmetic	0.35	−0.35
Algebra	0.29	−0.22
Geometry	0.17	−0.10

Unlike principal component score coefficients, factor score coefficients are not simple multiples of the loadings, so their interpretation differs from that of the factor loadings. The factor score coefficients allow the scores to be calculated from the original (standardized) variables. Examination of the coefficients is most relevant when we are concerned with individuals and wish to understand what their scores mean. The commoner practice of examining the factor loadings is more relevant when our interest is in the population structure and the interrelationships between the variables. Sometimes one factor score coefficient is dominant, so that the scores are virtually proportional to that variable. In such a case, anyone using the factor scores should realise that they are based almost entirely on a single variable.

6.8 A worked example: the test anxiety inventory

The following example illustrates a typical factor analysis, including the use of oblique rotation. This example is based on data from the test anxiety inventory, which is used to assess overall anxiety associated with taking tests. The inventory has been used in many countries, with similar results. The data analyzed here are from a study by Gierl and Rogers (1996) who applied a test with 20 items to 335 male and 389 female grade 12 students in British Columbia. A factor analysis of these data is also discussed in Bartholomew and Knott (1999). The analysis of the male sample is presented here.

Students were asked to report how frequently they experienced various symptoms of anxiety in taking tests. A brief description of the items is given below.

1. Lack of confidence during tests
2. Uneasy, upset feeling

3. Thinking about grades

4. Freeze up

5. Thinking about getting through school

6. The harder I work, the more confused I get

7. Thoughts interfere with concentration

8. Jittery when taking tests

9. Even when prepared, get nervous

10. Uneasy before getting the test back

11. Tense during test

12. Exams bother me

13. Tense/ stomach upset

14. Defeat myself during tests

15. Panicky during tests

16. Worry before important tests

17. Think about failing

18. Heart beating fast during tests

19. Can't stop worrying

20. Nervous during test, forget facts

The question of how, precisely, students were scored on these items would be relevant for judging the suitability of making the normality assumption but not for our more limited object here.

The correlation matrix is given in Table 6.9, and this shows that pairs of responses are positively correlated to a modest degree. This suggests that there is at least one common factor underlying the scores, which is what one would expect when the items have been specifically constructed to reflect test anxiety. However, it is not clear whether one factor is sufficient to account for the observed correlations, and this is where we might hope that factor analysis will reveal something further. In order to get some idea of the number of factors likely to be necessary, we first carry out a PCA. The result, given in Table 6.10, shows that there is one dominant eigenvalue and a second that is greater than one. These two together account for about half of the variance. Beyond the second eigenvalue, the rate of decrease is very slow. It seems clear that at least two factors are needed.

The unrotated factor loadings from a two-factor model are given in Table 6.11: x_1 corresponds to the first item, x_2 to the second item, and so on. A plot of these loadings is given in Figure 6.3. The first factor is positively correlated with each item and may be interpreted as a measure of overall anxiety during exams. However, the interpretation of the second factor is not immediately clear. It is obvious from Figure 6.3 that no orthogonal rotation will result in simple structure.

In an attempt to clarify the interpretation of the factors, an oblique rotation was carried out, and the new oblique axes (transformed factors) have been added to Figure 6.3. The rotated factor loadings (the *pattern matrix*) are shown in Table 6.11. While the loadings do not show simple structure, they

Table 6.9 *Pairwise correlations between test anxiety inventory items*

	x_1	x_2	x_3	x_4	x_5	x_6	x_7	x_8	x_9	x_{10}
x_1	1.00	.51	.24	.42	.31	.29	.39	.43	.43	.26
x_2	.51	1.00	.30	.41	.23	.34	.39	.48	.44	.28
x_3	.24	.30	1.00	.37	.40	.27	.54	.38	.34	.26
x_4	.42	.41	.37	1.00	.35	.34	.42	.41	.42	.30
x_5	.31	.23	.40	.35	1.00	.34	.49	.30	.22	.19
x_6	.29	.34	.27	.34	.34	1.00	.39	.26	.27	.25
x_7	.39	.39	.54	.42	.49	.39	1.00	.45	.41	.32
x_8	.43	.48	.38	.41	.30	.26	.45	1.00	.54	.33
x_9	.43	.44	.34	.42	.22	.27	.41	.54	1.00	.36
x_{10}	.26	.28	.26	.30	.19	.25	.32	.33	.36	1.00
x_{11}	.50	.47	.33	.39	.34	.35	.45	.59	.53	.39
x_{12}	.39	.41	.28	.42	.26	.33	.43	.48	.42	.33
x_{13}	.40	.43	.29	.35	.31	.28	.38	.43	.37	.29
x_{14}	.41	.37	.35	.49	.43	.42	.48	.34	.36	.27
x_{15}	.52	.49	.40	.60	.34	.36	.49	.56	.52	.39
x_{16}	.47	.48	.38	.44	.27	.31	.41	.53	.55	.43
x_{17}	.40	.30	.43	.46	.55	.37	.51	.36	.35	.32
x_{18}	.36	.46	.35	.41	.30	.27	.41	.48	.48	.35
x_{19}	.39	.34	.38	.36	.26	.23	.44	.45	.43	.42
x_{20}	.38	.33	.29	.53	.30	.40	.45	.44	.41	.30

	x_{11}	x_{12}	x_{13}	x_{14}	x_{15}	x_{16}	x_{17}	x_{18}	x_{19}	x_{20}
x_1	.50	.39	.40	.41	.52	.47	.40	.36	.39	.38
x_2	.47	.41	.43	.37	.49	.48	.30	.46	.34	.33
x_3	.33	.28	.29	.35	.40	.38	.43	.35	.38	.29
x_4	.39	.42	.35	.49	.60	.44	.46	.41	.36	.53
x_5	.34	.26	.31	.43	.34	.27	.55	.30	.26	.30
x_6	.35	.33	.28	.42	.36	.31	.37	.27	.23	.40
x_7	.45	.43	.38	.48	.49	.41	.51	.41	.44	.45
x_8	.59	.48	.43	.34	.56	.53	.36	.48	.45	.44
x_9	.53	.42	.37	.36	.52	.55	.35	.48	.43	.41
x_{10}	.39	.33	.29	.27	.39	.43	.32	.35	.42	.30
x_{11}	1.00	.53	.47	.45	.55	.60	.38	.54	.45	.45
x_{12}	.53	1.00	.32	.41	.52	.53	.41	.43	.38	.51
x_{13}	.47	.32	1.00	.41	.44	.40	.31	.50	.46	.35
x_{14}	.45	.41	.41	1.00	.52	.44	.40	.39	.38	.50
x_{15}	.55	.52	.44	.52	1.00	.65	.52	.52	.55	.56
x_{16}	.60	.53	.40	.44	.65	1.00	.45	.51	.54	.49
x_{17}	.38	.41	.31	.40	.52	.45	1.00	.35	.46	.40
x_{18}	.54	.43	.50	.39	.52	.51	.35	1.00	.48	.43
x_{19}	.45	.38	.46	.38	.55	.54	.46	.48	1.00	.46
x_{20}	.45	.51	.35	.50	.56	.49	.40	.43	.46	1.00

Table 6.10 *Variance explained by each principal component, test anxiety inventory items*

Component	Variance	Variance explained %	Cumulative %
1	8.78	43.90	43.90
2	1.35	6.75	50.65
3	0.97	4.86	55.51
4	0.89	4.44	59.95
5	0.77	3.87	63.82
6	0.74	3.71	67.53
7	0.71	3.53	71.06
8	0.66	3.27	74.33
9	0.57	2.85	77.18
10	0.54	2.72	79.90
11	0.54	2.71	82.61
12	0.51	2.53	85.14
13	0.47	2.36	87.50
14	0.44	2.18	89.68
15	0.42	2.10	91.78
16	0.38	1.89	93.67
17	0.35	1.74	95.41
18	0.33	1.66	97.07
19	0.31	1.57	98.64
20	0.28	1.38	100.00

do have an interpretable pattern. Factor 1 has high positive loadings on variables $x_1, x_2, x_4, x_8, \ldots, x_{13}, x_{15}, x_{16}$, and x_{18}, x_{19}, x_{20}, while factor 2 has high positive loadings on the remaining set of variables. The items that have the highest correlations with factor 1 are largely measures of "emotionality", that is, reactions evoked by the nervous system. Original studies have identified items $x_2, x_8, x_9, x_{10}, x_{15}, x_{16}$, and x_{18} as the "emotionality" items. The items that have the highest correlations with factor 2 can be thought of as measures of a more psychological type of anxiety which we might call "worry". Items $x_3, x_4, x_5, x_6, x_7, x_{14}, x_{17}$, and x_{20} are identified as indicators of "worry". Although the points do not lie along one or other axis, they do fall roughly into two groups.

After an oblique rotation, the factors will be correlated. The correlation between the "worry" and "emotionality" factors from the analysis of the test anxiety inventory items is estimated as 0.684, indicating that the two factors are quite closely related. Because the factors are themselves correlated, the rotated factor loadings now given in the pattern matrix are no longer the correlations between the observed variables and the factors. These correlations are given separately in the *structure matrix*. The structure matrix from the oblique rotation of the two-factor model of the test anxiety inventory data is given in Table 6.12. All the variables are positively correlated with both factors, which means that each item measures, to some extent, both the "worry" factor and the "emotionality" factor.

Table 6.11 *The unrotated and OBLIMIN rotated loadings from a two-factor model of the test anxiety inventory items, males*

| | Unrotated | | Rotated | |
	$\hat{\alpha}_{i1}$	$\hat{\alpha}_{i2}$	$\hat{\alpha}_{i1}^*$	$\hat{\alpha}_{i2}^*$
x_1	0.62	−0.07	0.56	0.09
x_2	0.62	−0.16	0.67	−0.04
x_3	0.54	0.25	0.14	0.49
x_4	0.65	0.09	0.41	0.31
x_5	0.51	0.50	−0.16	0.82
x_6	0.49	0.20	0.16	0.41
x_7	0.67	0.29	0.21	0.58
x_8	0.69	−0.18	0.75	−0.04
x_9	0.66	−0.21	0.76	−0.10
x_{10}	0.50	−0.09	0.49	0.02
x_{11}	0.73	−0.18	0.78	−0.03
x_{12}	0.65	−0.10	0.62	0.05
x_{13}	0.59	−0.06	0.53	0.09
x_{14}	0.64	0.18	0.30	0.42
x_{15}	0.80	−0.07	0.71	0.13
x_{16}	0.75	−0.21	0.83	−0.07
x_{17}	0.64	0.34	0.12	0.64
x_{18}	0.67	−0.14	0.68	0.01
x_{19}	0.65	−0.06	0.59	0.10
x_{20}	0.66	0.02	0.50	0.21

Table 6.12 *Structure matrix giving correlations between test anxiety inventory items and the two rotated factors after an OBLIMIN rotation, males*

| | Factor | |
	1	2
x_1	0.62	0.47
x_2	0.64	0.41
x_3	0.48	0.59
x_4	0.62	0.59
x_5	0.40	0.71
x_6	0.44	0.52
x_7	0.60	0.72
x_8	0.72	0.47
x_9	0.69	0.42
x_{10}	0.50	0.35
x_{11}	0.75	0.50
x_{12}	0.66	0.50
x_{13}	0.59	0.45
x_{14}	0.59	0.63
x_{15}	0.80	0.61
x_{16}	0.78	0.50
x_{17}	0.56	0.72
x_{18}	0.68	0.47
x_{19}	0.65	0.50
x_{20}	0.64	0.55

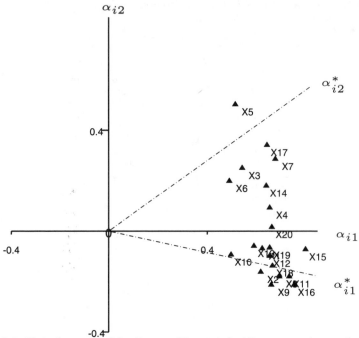

Figure 6.3 *Plot of unrotated loadings, with rotated oblique axes shown, for a two-factor model of the test anxiety inventory items, males*

6.9 How rotation helps interpretation

Now that we have seen the process of rotation in action, it is worth looking again at the issues raised by having an infinite number of solutions available when there are two or more factors. All estimation routines are designed to yield what we might call a "standard" solution which is similar to the PCA solution. Often, this solution lends itself to direct interpretation as in the case of the example of the subject marks considered earlier and other examples in Section 6.11 below. In other cases, rotation may lead to a clearer interpretation. We argued earlier that rotation gives us alternative ways of describing the same solution rather than providing different solutions. The fundamental thing about the solution, in this example, is that it takes two dimensions to describe the latent variation between individuals. Or, put negatively, one dimension is not sufficient to describe the variation among individuals in their responses to these test items as, one supposes, the original designers of the test had intended.

The standard solution, like PCA, produced a dominant factor with fairly large and positive loadings on all variables. This indicated that there is something common to all test items which we may equate with the "test anxiety" which the test was designed to measure. But it also showed that, even among those at the same point on this scale, there was some further variation which we could not readily interpret. Rotation enabled us to look at the variation from another angle. It proved possible to describe it in terms of what earlier researchers have called worry and emotionality. Although these two dimensions are highly correlated they appear to be distinct aspects of what we usually call "anxiety". The analysis shows us that it is compounded of something which has a physiological origin in the nervous system and something which seems to be more directly psychological. In retrospect, this may not seem particularly surprising but the distinction does not appear to have been recognised until it was revealed by factor analysis (see Gierl and Rogers (1996) and references therein).

6.10 A comparison of factor analysis and principal components analysis

PCA and factor analysis share some aims. Both methods attempt to reduce the dimensionality of a set of correlated variables, $x_1, \ldots, x_p$, by obtaining a small set of components or factors, $y_1, \ldots, y_q$. PCA finds components that account for as much as possible of the total variance, $\sum_i \text{var}(x_i)$, whereas FA tries to match the reconstructed correlations to the observed sample correlations.

Neither method is useful if the xs have low correlations; in PCA, the components will be close to the original variables, and in FA there will be little correlation to explain. Although the methods have common aims, however, the procedures by which they achieve these aims are rather different.

i) *Descriptive versus model-based.* PCA is a descriptive technique which does not assume an underlying statistical model. The principal components are simply transformed versions of the xs. It makes no prior assumptions about how many components are being looked for or what they might represent. The most that is hoped for is that a few ys will provide a good summary of the observed xs for the given sample. Factor analysis assumes a statistical model which incorporates a fixed number of factors and there may be some prior notion of what they represent. This means that factor analysis may be used to make inferences about the population from which the sample was drawn. For example in confirmatory factor analysis, we might test whether the factor loadings follow a pre-specified pattern.

Confirmatory factor analysis and structural equation modelling are powerful tools for social science research, building on the ideas of regression, path analysis and factor analysis. Unfortunately, they are open to misuse. Correct interpretation of the results is not always easy. While these more complicated techniques are beyond the scope of this book, we hope that by developing a basic understanding of FA the reader will be better prepared to learn how and when to use them.

ii) *Rotation.* The first principal component is always the same, no matter how many further components are extracted. But the single factor for a one-factor model will not be the same as the first factor in a two-factor model. Thus, if we choose to rotate principal components, for example in the search for simple structure, we lose some information. With FA, all solutions for the q-factor model are equivalent, defining the same q-dimensional sub-space in which the correlation structure can best be represented. Rotating a solution for FA neither adds nor loses information, but it may help interpretation.

iii) *Score coefficients.* Often the component loadings for PCA and for FA will be very similar when both analyses are performed on a data set with strong correlation structure. However in converting to score coefficients, differences may arise. PCA aims to explain the variation in all the variables and the score coefficients are simple multiples of the loadings:

$$\tilde{a}_{ij} = \lambda_j a_{ij}^* \quad (i = 1, \ldots, p; \; j = 1, \ldots, q).$$

FA, on the other hand, only aims to explain the correlations, so variables with small loadings (correlations) with a factor will contribute even less to the factor scores. The larger loadings are multiplied by a larger amount and the differences between the factor loadings are accentuated. As already mentioned in Section 6.7, and as illustrated in Section 6.11 in the extreme case, the scores for a factor may be almost a linear function of a single variable with factor score coefficients for other variables close to zero.

A formal comparison of PCA and factor analysis

In spite of these important differences, there are circumstances under which PCA is a good approximation to factor analysis, and it is this fact which provides the basis for using PCA as a guide to choosing the number of factors. As outlined in Section 5.8, in PCA, a set of p uncorrelated components, each with unit variance, can be derived from the original p (standardized) variables:

$$\tilde{y}_j = \tilde{a}_{j1} x_1 + \tilde{a}_{j2} x_2 + \cdots + \tilde{a}_{jp} x_p \quad (j = 1, \ldots, p).$$

We can invert these equations to obtain a set of p equations with the xs on the left hand side, and the ys on the right hand side giving:

$$x_i = a_{i1}^* \tilde{y}_1 + a_{i2}^* \tilde{y}_2 + \cdots + a_{ip}^* \tilde{y}_p \quad (i = 1, \ldots, p),$$

where the a^*s are the loadings (or correlations) with the principal components. Suppose we retain only the first q components, then we obtain:

$$x_i = a_{i1}^* \tilde{y}_1 + a_{i2}^* \tilde{y}_2 + \cdots + a_{iq}^* \tilde{y}_q + u_i \quad (i = 1, \ldots, p).$$

This equation now has the same form as the factor analysis model, that is, each x is expressed as a linear combination of q uncorrelated variables, each with variance one. But for this to be close to a factor analysis model, the residual terms, u_i, would need to behave like the uncorrelated residuals or specific factors, e_i, in the factor model. In general, the u_i will not be uncorrelated. However, under some circumstances, the two analyses will give similar results, for example if the variances of the e_is in the factor model, σ_i^2, are roughly equal or if the σ_i^2s are all small. The reader is recommended to compare the results of PCA and FA carried out on the same datasets.

6.11 Further examples and suggestions for further work

Psychomotor tests

A sample of 197 airmen was subjected to a range of tests (Fleishman and Hempel 1954). A subset of these tests has been selected for analysis and brief descriptions of those selected are given in Table 6.13. Further details can be found in Fleishman and Hempel (1954). The first of these tests was a criterion practice task in which the airmen's performance was assessed in repeated trials over a two-day period. The practice period was divided into eight time segments and the scores in each were obtained to give variables x_1 to x_8. The remaining tests were of two types: written tests which aimed to assess speed in performing verbal, spatial, and arithmetic tasks (x_9, x_{10}, and x_{11}), and practical tests to assess speed and accuracy in operating apparatus and reaction times to various stimuli (x_{12}, x_{13}, and x_{14}).

Table 6.13 *Descriptions of psychomotor test items*

Type of test	Test	Variable
Criterion task: complex coordination	Stage 1	x_1
Criterion task: complex coordination	Stage 2	x_2
Criterion task: complex coordination	Stage 3	x_3
Criterion task: complex coordination	Stage 4	x_4
Criterion task: complex coordination	Stage 5	x_5
Criterion task: complex coordination	Stage 6	x_6
Criterion task: complex coordination	Stage 7	x_7
Criterion task: complex coordination	Stage 8	x_8
Printed tests of comprehension and speed	Numerical operations	x_9
Printed tests of comprehension and speed	Dial and table reading	x_{10}
Printed tests of comprehension and speed	Mechanical principles	x_{11}
Apparatus tests	Plane control	x_{12}
Apparatus tests	Reaction time	x_{13}
Apparatus tests	Rate of movement	x_{14}

The pairwise correlations between the 14 test scores are given in Table 6.14. As would be expected, you can see some very high correlations between the criterion test scores over the eight time segments (x_1 to x_8), particularly between scores taken for segments that are close together in time. Among the printed tests, there are quite high correlations between x_9 and x_{10}, and between x_{10} and x_{11}, while the correlations among the apparatus tests are low to moderate.

Before carrying out a factor analysis of these data, you should carry out a principal components analysis. The scree plot from a PCA is shown in Figure 6.4. You can see that the first component is highly dominant. From a PCA, you will find that the first component explains 54% of the total variance, while the second and third explain 12 and 8%, respectively. Only the first

Table 6.14 *Pairwise correlations (× 100) between psychomotor test items*

	x_1	x_2	x_3	x_4	x_5	x_6	x_7	x_8	x_9	x_{10}	x_{11}	x_{12}	x_{13}	x_{14}
x_1	100	75	73	66	64	57	63	59	28	51	49	40	08	25
x_2	75	100	85	85	84	79	77	79	30	46	40	45	22	32
x_3	73	85	100	85	83	79	81	79	30	45	39	44	27	31
x_4	66	85	85	100	90	88	86	85	26	40	36	44	30	28
x_5	64	84	83	90	100	90	87	86	22	37	36	42	30	34
x_6	57	79	79	88	90	100	85	86	23	34	29	39	27	37
x_7	63	77	81	86	87	85	100	90	23	36	33	39	33	30
x_8	59	79	79	85	86	86	90	100	24	34	30	36	27	32
x_9	28	30	30	26	22	23	23	24	100	63	32	8	9	12
x_{10}	51	46	45	40	37	34	36	34	63	100	54	22	5	24
x_{11}	49	40	39	36	36	29	33	30	32	54	100	22	-5	12
x_{12}	40	45	44	44	42	39	39	36	8	22	22	100	20	20
x_{13}	8	22	27	30	30	27	33	27	9	5	−5	20	100	30
x_{14}	25	32	31	28	34	37	30	32	12	24	12	20	30	100

three eigenvalues are greater than one. The results from the PCA suggest that a two- or three-factor model should be fitted. Here, we examine only the two-factor solution, but you should also consider the three-factor model.

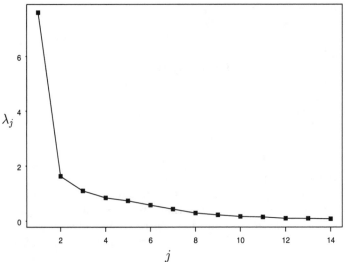

Figure 6.4 *Scree plot of eigenvalue versus number of component from principal components analysis of the psychomotor test data*

Table 6.15 shows the communalities obtained from the two-factor model. These are very high for seven of the criterion task variables (x_2 to x_8) and

Table 6.15 *Communalities from fitting a two-factor model to the psychomotor test data*

x_1	x_2	x_3	x_4	x_5	x_6	x_7	x_8	x_9	x_{10}	x_{11}	x_{12}	x_{13}	x_{14}
0.64	0.82	0.82	0.89	0.90	0.87	0.85	0.84	0.35	0.64	0.40	0.21	0.12	0.12

high for x_1 and x_{10}. In contrast, the variances of the apparatus test scores are not well explained by two common factors. You might like to see whether adding a third factor leads to higher communalities for these variables. The log-likelihood-ratio test statistic for overall goodness-of-fit also indicates that the two-factor model is not a good fit ($W = 184.2$, degrees of freedom=64, $p < 0.001$). As a further check on fit, you should examine the reproduced correlation matrix.

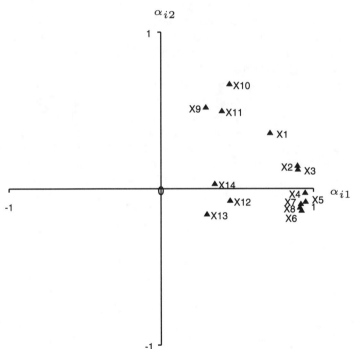

Figure 6.5 *Plot of loadings from a two-factor model of the psychomotor test data*

Although the two-factor model does not appear to be a good fit, it may still reveal key patterns in the data. Figure 6.5 shows a plot of the factor loadings from the two-factor model. From this plot, you can see that the tests are roughly separated into three groups. The printed tests of comprehension and speed (x_9, x_{10}, and x_{11}) all have moderate, positive loadings on the first factor and high, positive loadings on the second factor. The apparatus tests (x_{12}, x_{13}, and x_{14}) form another cluster with positive, moderate loadings on the first factor and close-to-zero loadings on the second factor. Seven of the criterion variables (x_2 to x_7) constitute a third cluster with high loadings on factor 1 and small contributions from factor 2. The first criterion variable, x_1, appears between the cluster of the remaining criterion variables and the cluster of the printed tests. You can also see that the criterion variables are approximately ordered on the second factor, with x_1 having the highest loading and the scores at the end of the practice period (x_6, x_7, and x_8) the lowest loadings. This might reflect a learning effect on the criterion task; in the first segment when the task is new, the airmen's performance (x_1) is correlated with their performance on (x_9, x_{10}, x_{11}), but with practice, their performance on the criterion task becomes less correlated with their scores on the printed tests.

Social mobility

In Section 5.9, a principal components analysis was carried out of a correlation matrix of variables relating to the occupational and educational status of three generations of family members (Table 5.14). A description of the ten variables used in the analysis was given in Table 5.13. The correlation matrix may also be analyzed using factor analysis.

A PCA of these data found that three, possibly four, components were needed and the first four components were all interpretable. This suggests that we should begin with a three-factor model. The log-likelihood-ratio test indicates that the three-factor solution is not a good fit to the correlation matrix ($W = 143.8$, degrees of freedom=18, $p < 0.001$). Since the fourth component from PCA was interpretable, the next step might be to fit a four-factor model. If you do this, you will find that four-factor model is a good fit ($W = 16.6$) degrees of freedom= 11, $p=0.12$). However, there is a problem with this model. Most software packages will inform you that during the fitting process the estimates of one or more communalities exceeded a value of 1. As the communality is a proportion, a value greater than 1 is clearly not permissable and therefore a solution where such values have occurred should be treated with caution. This phenomenon is known as a *Heywood case*. If you examine the communalities from the four-factor solution, you will find that the value for x_6 is extremely close to 1 — an indication of a Heywood case. One solution to the problem might be to omit x_6 from the analysis, but if you try this, you will find that the problem is merely shifted elsewhere as the communality of another variable is estimated very close to 1. In such cases, it is recommended that a factor be dropped. We therefore return to the three-factor solution even though it is a much poorer fit than the four-factor solution.

The factor loadings obtained from the three-factor model are given in Table 6.16. If you examine the pattern of loadings for each factor, you will find that the interpretation of the factors is the same as the interpretation of the first three principal components. To aid interpretation, relatively large loadings on the second and third factors are printed in bold. You can see that the first factor is positively correlated with all ten variables; the second factor contrasts occupational status of family members (x_1, x_2, x_5) with qualifications (x_4, x_7, x_9); while the third factor contrasts variables relating to the first born son (x_8, x_9, x_{10}) with the educational status of his mother (x_6, x_7).

Table 6.16 *Loading matrix giving the unrotated loadings from a three-factor model of the social mobility data*

		$\hat{\alpha}_{i1}$	$\hat{\alpha}_{i2}$	$\hat{\alpha}_{i3}$
x_1	HF/O	0.426	**0.403**	0.053
x_2	WF/O	0.404	**0.343**	0.008
x_3	H/FE	0.592	−0.026	0.116
x_4	H/Q	0.558	**−0.240**	0.118
x_5	H/O	0.575	**0.481**	0.031
x_6	W/FE	0.451	−0.126	**0.369**
x_7	W/Q	0.477	**−0.296**	**0.462**
x_8	FB/FE	0.615	−0.191	**−0.289**
x_9	FB/Q	0.519	**−0.358**	**−0.381**
x_{10}	FB/O	0.602	0.168	**−0.219**

Table 6.17 *Loading matrices giving the VARIMAX and OBLIMIN rotated loadings from a three-factor model of the social mobility data*

		VARIMAX			OBLIMIN		
		$\hat{\alpha}^*_{i1}$	$\hat{\alpha}^*_{i2}$	$\hat{\alpha}^*_{i3}$	$\hat{\alpha}^*_{i1}$	$\hat{\alpha}^*_{i2}$	$\hat{\alpha}^*_{i3}$
x_1	HF/O	**0.576**	0.042	0.111	−0.064	**0.599**	0.025
x_2	WF/O	**0.516**	0.086	0.090	−0.003	**0.530**	0.002
x_3	H/FE	0.329	0.288	**0.416**	0.183	0.246	**0.353**
x_4	H/Q	0.135	0.360	**0.485**	0.279	0.015	**0.445**
x_5	H/O	**0.728**	0.113	0.144	−0.016	**0.747**	0.025
x_6	W/FE	0.163	0.078	**0.568**	−0.051	0.074	**0.585**
x_7	W/Q	0.042	0.106	**0.718**	−0.032	−0.085	**0.765**
x_8	FB/FE	0.209	**0.645**	0.194	**0.637**	0.101	0.058
x_9	FB/Q	0.018	**0.723**	0.140	**0.762**	−0.109	0.014
x_{10}	FB/O	**0.491**	**0.434**	0.098	**0.381**	**0.452**	−0.052

Although the unrotated factors are interpretable, rotations can be carried out to determine whether simple structure can be achieved. The factor loadings obtained from an orthogonal (VARIMAX) rotation and an oblique (OBLIMIN) rotation of the three-factor solution are shown in Table 6.17. Again, relatively large loadings are printed in bold. A VARIMAX rotation has not led to a simple structure since there are several variables which have moderate loadings on more than one factor. However, the interpretation of the factors has been made clearer. The first factor has large loadings on the occupation variables and might be labelled "occupational status", while the second and third factors might be labelled "first born son" and "parents' education", respectively.

A pattern of loadings that is close to simple structure is achieved by allowing the factors to correlate via an OBLIMIN rotation. An oblique rotation seems reasonable in this case since, for example, you would expect a son's occupational and educational status to be positively correlated with that of his parents. If you look at the correlations between the OBLIMIN rotated factors you will find that they are all positive, ranging from 0.35 to 0.40. However, an oblique rotation does not change the overall interpretation of the solution.

6.12 Further reading

Bartholomew, D. J. and Knott, M. (1999). *Latent Variable Models and Factor Analysis*. 2nd edition. London: Arnold.

Basilevsky, A. (1994). *Statistical Factor Analysis and Related Methods*. New York: Wiley.

CHAPTER 7

Factor Analysis for Binary Data

7.1 Latent trait models

In this chapter, we move to the top right hand cell of Table 6.1 and discuss methods based on models where the manifest variables are categorical. We start with the case where they are all binary — that is, where they are based on responses of the kind yes/no or right/wrong. Some methods appropriate when there are more than two categories will be given in Chapter 8. The word "trait" in the name of these models is often used because it arises from one of the principal applications for which they were devised, namely the measurement of psychological traits. In this book, they are used in a much broader context and so it seemed appropriate to make this clear in the title of the chapter. Nevertheless, we have also retained the original terminology to keep the link with a very important field of application.

Conceptually there is no difference between the problems treated here and those in the previous chapter on factor analysis. We start with a probability model linking the observed variables to a set of latent variables. We then discuss how to fit the models, judge their goodness-of-fit, interpret their parameters, and so forth. The difference lies in the special problems posed by having to deal with a data matrix consisting of binary items. The basic objectives are the same, namely:

i) To explore the interrelationships between the observed responses

ii) To determine whether the interrelationships can be explained by a small number of latent variables

iii) To assign a score to each object for each latent variable on the basis of its responses

The binary data matrix

We have already met a data matrix for categorical data in the discussions of cluster analysis, multidimensional scaling and, in passing, correspondence analysis. If the responses are binary, the xs simply record whether the response was positive or negative. A convenient convention, also used in earlier chapters, is to use 1 to indicate a "success" or a positive response, that is "correct" or "yes" as the case may be, and 0 for the "failure" or negative response. This convention has the advantage in the present context that if we sum the responses in any row of the data matrix, we get the total number of positive responses. This is a useful summary measure in its own right and we shall use it in the subsequent analyses. The response coded 1 is sometimes referred to as the *keyed response*. A typical row of the data matrix might then be as follows:

$$00101110011$$

The methods about to be described all start from a data matrix consisting of a set of rows like that above, one for each individual or object. However, the restriction to binary data sometimes makes it possible to express the matrix in a more compact and informative way.

Any row of the data matrix is referred to as a *score pattern* or a *response pattern*. If there are p variables there are 2^p possible response patterns. When $p = 3$, for example, they are

$$000, \ 001, \ 010, \ 011, \ 100, \ 101, \ 110, \ 111.$$

If the sample size is much larger than 2^p, many of the response patterns will be repeated. It is, therefore, much more economical to present the matrix as a list of the possible response patterns together with their associated frequencies as follows:

000	175
001	64
010	17
100	12
011	9
101	3
110	33
111	98

The second column records how many times each response pattern occurs. This grouped form of the data matrix is used whenever the sample size is large. However, when the number of variables p is large, many response patterns may not occur at all, in which case they are omitted from the table to save space.

Latent trait methods were introduced in educational testing where most of their development has taken place; this is now a highly specialised field with a substantial literature of its own. Our emphasis in this chapter will be mainly on their general use as tools for social research in the factor analysis tradition.

An example

To illustrate the various steps in the analysis, we shall use a data set with only four variables extracted from the 1986 British Social Attitudes Survey (McGrath and Waterton, 1986). The data are the responses given by 410 individuals to four out of seven items concerning attitude to abortion. A small proportion of non-response occurs for each item, the proportions being (0.03, 0.03, 0.05, 0.04). In order to avoid the distraction of having to deal with missing values, we have slightly adjusted the data to eliminate missing values. An analysis that includes all respondents and uses a factor analysis (FA) model for binary items that takes account of missing values was carried out by Knott, Albanese, and Galbraith (1990) . The results were not substantially different from those reported here. After eliminating the missing values, we are left with 379 respondents. For each item, respondents were asked if the law should allow abortion under the circumstances presented under each item. The four items used in the analysis are given below:

1. The woman decides on her own that she does not. [WomanDecide]
2. The couple agree that they do not wish to have the child. [CoupleDecide]
3. The woman is not married and does not wish to marry the man. [NotMarried]
4. The couple cannot afford any more children. [CannotAfford]

The frequency of each response pattern is given in Table 7.1.

Table 7.1 *Frequencies of response patterns, attitude towards abortion*

Response patterns	Frequency
1111	141
0000	103
0111	44
0011	21
0001	13
1110	12
0010	10
0100	9
0110	7
1011	6
0101	6
1101	3
1100	3
1000	1
1010	0
1001	0
Total	379

We find that the percentage of individuals agreeing that abortion should be legal under circumstances described by the items 1 to 4 are 43.8, 59.4, 63.6, and 61.7%, respectively. If we were doing a factor analysis, we would next compute the correlations between pairs of variables and inspect the result, looking for evidence of positive correlations which suggest that there might be one or more common underlying factors. In the case of binary data, the corresponding things to look at are the pairwise associations between variables. We can do this by constructing 2×2 contingency tables. For example, Table 7.2 cross-tabulates the first two items that show a strong association. A similar analysis for other pairs of variables produces similar results. This suggests that it would be worth asking whether these associations can be attributed to one or more

Table 7.2 *Cross tabulation of items 1 and 2, attitude towards abortion*

	Yes	No
Yes	159	7
No	66	147

common factors. This is what a latent trait model enables us to do. If we can identify common factors, we may then wish to go on to compute scores for individuals on the latent dimensions.

7.2 Why is the factor analysis model for metrical variables invalid for binary responses?

Since the approach for binary and metrical variables has been so similar up to this point, it is natural to think of treating the binary data as if they were metrical. What is to prevent us from computing the product moment correlations and doing a factor analysis in the usual way? There is no practical bar to doing just that, and one sometimes finds such factor analyses in the research literature. However, such an analysis is inappropriate because it is based on a model which assumes that the observed or manifest variables $(x_1, \ldots, x_p)$ are metrical rather than binary. To see why this is so, we briefly return to the factor analysis model. The model was written as:

$$x_i = \alpha_{i0} + \alpha_{i1} y_1 + \cdots + \alpha_{iq} y_q + e_i \qquad (i = 1, \ldots, p), \qquad (7.1)$$

where p denotes the total number of observed items, x_i denotes the ith metrical observed item, $\mathbf{y} = (y_1, \ldots, y_q)$ denotes the vector of latent variables and e_i denotes the residual. We assume that the residual follows a normal distribution with mean 0 and variance σ_i^2, the latent variables are assumed to be independent with standard normal distributions $y_j \sim N(0,1)$ for all j. Since $\mathbf{y}$ and e_i can take any value and are independent of each other, x_i can also take any value. Therefore, the linear factor model is invalid for categorical variables in general and for binary variables in particular.

We need a different model to relate the latent variables $\mathbf{y}$ to the manifest variables. Two approaches have been adopted to meet this need. The oldest is to try to retain as much as possible of the factor analysis method. This is done by imagining a fictitious variable for each i which is partially revealed to us by x_i. This enables us to retain the factor model for the (unobserved) fictional variable. This method is still widely used and we shall describe it in Section 7.7.

A better approach is to start, as we did in factor analysis, with the idea of a regression model. We want an appropriate model for the regression of each x_i on the latent variables. The usual regression method used for an observable binary response on a set of observable explanatory variables is known as *logistic* regression. It takes its name from the logistic function used in the regression equation.

In order to motivate the choice of this function, we first remind ourselves that the regression of x_i on the latent variables is the expected value of x_i given the ys. Since x_i is binary, the expected value of x_i given the ys is the same as $\Pr(x_i = 1 \mid \mathbf{y}) = \pi_i(\mathbf{y})$ where $\pi_i(\mathbf{y})$ is the conditional probability that binary variable, x_i, equals one given the values of the q latent variables $y_1, \ldots, y_q$. We, therefore, have to specify the form of the probability $\pi_i(\mathbf{y})$ as a function of $y_1, \ldots, y_q$. The function chosen is known as the *link* function.

An identical linear link function would be the simplest giving:

$$\pi_i(\mathbf{y}) = \alpha_{i0} + \alpha_{i1}y_1 + \cdots + \alpha_{iq}y_q \qquad (i = 1, \ldots, p). \tag{7.2}$$

But such a linear relationship between the probability of a correct response and the latent variables has two flaws.

i) The left hand side of equation (7.2) is a probability that takes values between 0 and 1, and the right hand side is not restricted in any way and can take any real value.

ii) We might expect that the rate of change in the probability of a correct/positive response will not be the same for the whole range of $\mathbf{y} = (y_1, \ldots, y_q)$. In that case, a curvilinear relationship might be more appropriate.

To take into account both those points, we need to introduce a different link function between the probability and the latent variables. That link should map the range $[0, 1]$ onto the range $(-\infty, +\infty)$. It should also be a monotonic function of each y because increasing any y should have the effect of increasing the probability. The logistic regression model uses the logit link illustrated in equation (7.3). We shall use the logit link mainly because it possesses theoretical and practical advantages (see Section 7.3). We shall also, in Section 7.7, briefly discuss the use of the normit link (also known as the probit) as an alternative when we consider the underlying variable (UV) approach.

7.3 Factor model for binary data

The logit model is defined as:

$$\mathrm{logit}\pi_i(\mathbf{y}) = \log_e \frac{\pi_i(\mathbf{y})}{1 - \pi_i(\mathbf{y})} = \alpha_{i0} + \sum_{j=1}^{q} \alpha_{ij}y_j. \tag{7.3}$$

By transforming $\pi_i(\mathbf{y})$ using the logit transformation, we have been able to write the model as linear in the latent variables which will greatly facilitate the interpretation. The probability $\pi_i(\mathbf{y})$ denotes the probability of "success" and the ratio $\pi_i(\mathbf{y})/(1 - \pi_i(\mathbf{y}))$ is also known as the odds of "success". We can rearrange equation (7.3) to get an expression for $\pi_i(\mathbf{y})$:

$$\pi_i(\mathbf{y}) = \frac{\exp(\alpha_{i0} + \sum_{j=1}^{q} \alpha_{ij}y_j)}{1 + \exp(\alpha_{i0} + \sum_{j=1}^{q} \alpha_{ij}y_j)}. \tag{7.4}$$

It may easily be checked that this expression behaves in the right way, namely that it lies between 0 and 1 and is monotonic in each y.

An important special case is obtained by putting $q = 1$. It is this case with which *item response analysis* is mainly concerned. Thus, we have the unidimensional latent trait model:

$$\pi_i(y_1) = \frac{\exp(\alpha_{i0} + \alpha_{i1}y_1)}{1 + \exp(\alpha_{i0} + \alpha_{i1}y_1)}.$$

In the psychometric literature, $\pi_i(y_1)$ is referred to as the item characteristic curve or item response function (IRF). It shows how the probability of a correct response increases with ability, say.

The logit model with one latent variable is plotted on Figure 7.1 for $\alpha_{i0} = 0.5$ and for different positive values of the parameter α_{i1} and on Figure 7.2 for different values of α_{i0} and for $\alpha_{i1} = 0.5$.

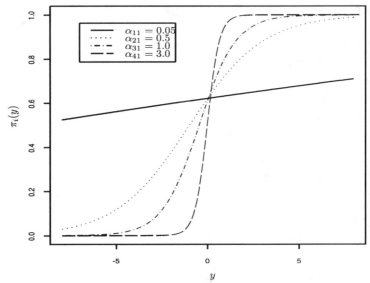

Figure 7.1 *Item characteristic curves for different values of the discrimination coefficient* α_{i1} *and* $\alpha_{i0} = 0.5$

It is clear that the parameter α_{i1} determines the steepness of the curve over the middle of the range. This means that a given change in the value of y_1 will produce a larger change in the probability of a positive response when this parameter is large than when it is small. For this reason, it is known in item response theory as the *discrimination* parameter. Increasing the parameter α_{i0} increases the probability for all values of y_1 and so it is referred to as the *difficulty* parameter.

We summarise and complete the specification of the factor model for binary data by listing the assumptions on which it depends as follows:

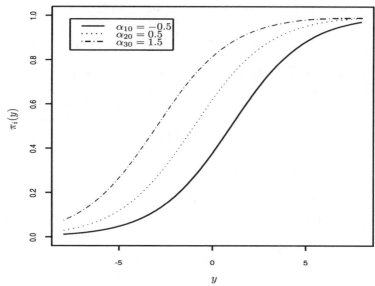

Figure 7.2 *Item characteristic curves for different values of the "difficulty" parameter α_{i0} and $\alpha_{i1} = 0.5$*

i) Conditional independence: the responses to the p observed items are independent conditional on the latent variables. In other words, the latent variables (factors) account for all the associations among the observed items. Since the latent variables are unobserved, the assumption of conditional independence can only be tested indirectly by checking whether the model fits the data. A latent variable model is accepted as a good fit when the latent variables account for most of the association among the observed responses.

ii) The link function: $\text{logit}\,\pi_i(\mathbf{y}) = \alpha_{i0} + \sum_{j=1}^{q} \alpha_{ij} y_j$, where $\Pr(x_i = 1 \mid \mathbf{y}) = \pi_i(\mathbf{y})$; $(i = 1, \ldots, p)$. A possible alternative would be the normit link, see Section 7.7, which gives very similar results in practice.

iii) The latent variables or factors $y_1, \ldots, y_q$ are independent with standard normal distributions. That is $y_j \sim N(0, 1)$ for $(j = 1, \ldots, q)$. The choice of the normal distribution for the latent variables has rotational advantages as we will see later but other distributions could be used. Fortunately, research has shown that the form of the distribution of the latent variables does not have much influence on the interpretation of the results of an analysis.

The Rasch model

A special case of the unidimensional model is obtained when all the discrimination parameters are equal ($\alpha_{11} = \alpha_{12} = \cdots = \alpha_{1p}$). This model was first discussed by Rasch (Rasch 1960) and it is usually written as:

$$\Pr(x_i = 1 \mid \alpha_{i0}, \beta_k) = \pi_{ik} = \frac{\exp(\alpha_{i0} + \beta_k)}{1 + \exp(\alpha_{i0} + \beta_k)}.$$

The random latent variable y and its coefficient have been replaced by the parameter β_k where $(k = 1, \ldots, n)$. This is because the model is usually used in educational testing to study a particular set of individuals, and it is their abilities that we are interested in. The β_ks measure each individual's ability, and the parameter α_{i0} remains the difficulty parameter. Nonetheless, we are more interested in the population from which our sample has been drawn and so we treat y as a variable with a probability distribution. The Rasch model is still quite popular in educational testing because of its simplicity and its attractive theoretical properties. In particular:

i) The total score $\sum_{i=1}^{p} x_{ik}$ is sufficient for β_k — that is, it contains all the information in the data about the βs if the model is true.

ii) The total number of positive/correct responses for item i, $\sum_{k=1}^{n} x_{ik}$ is sufficient for α_{i0}.

Fitting the logit model

Recall that in factor analysis, we fitted the model by choosing the parameter values to make the covariance matrix predicted by the model as close as possible to the observed matrix. For that model, the joint distribution was completely determined by the covariances so, in effect, we were making the observed and predicted distributions as close as possible. We do essentially the same thing when fitting the latent trait model. We choose the parameter values which make the frequency distribution across responses predicted by the model as close as possible to the observed one. As in factor analysis, there are various ways in which this distance can be measured but the one for which software is currently available is based on the likelihood function — the maximum likelihood method.

Interpretation of model parameters

In the latent trait model for each observed item i, we have $q + 1$ parameters to estimate, the intercept α_{i0} and the factor loadings $\alpha_{i1}, \ldots, \alpha_{iq}$. We have already noted that α_{i0} is called the difficulty parameter in educational testing because of its effect on the probability of a positive response. This effect can be seen more clearly if we consider the position when $\mathbf{y} = \mathbf{0}$. Since the ys are assumed to have standard normal distributions, an individual at this point in the latent space may be described as the "median" individual because, on each dimension, half the population lie on either side. In those circumstances we find,

$$\Pr(x_i = 1 \mid \mathbf{y} = \mathbf{0}) = \pi_i(\mathbf{y} = \mathbf{0}) = \frac{\exp(\alpha_{i0})}{1 + \exp(\alpha_{i0})}.$$

This is the probability that the median individual will respond correctly or positively to item i. For practical purposes, $\pi_i(\mathbf{0})$ is more directly interpretable than α_{i0}.

The α_{ij}s $(j = 1, \ldots, q)$ are factor loadings, but we have already noted that they are known in item response theory as discrimination parameters. The larger the value of α_{ij}, the greater is the effect of factor j on the probability of a positive response to item i; equivalently, the higher the value of α_{ij} for an item, the greater the difference in the probabilities of getting a correct/positive response between two individuals who are located at some distance apart on the latent dimensions. As a result, it will be easier to discriminate between those two individuals on the evidence of their responses to that item. The factor loadings α_{ij} are not bounded in any way, and for some items they may take very large values, indicating a very steep slope for the item response curve. This phenomenon is known as a "threshold effect", and we shall meet it again in Chapter 9. Large estimates of the discrimination parameters often have large standard errors, which means that their values are poorly determined. The maximum likelihood estimates for the attitude to abortion data are given in Table 7.3 along with their asymptotic (i.e., estimated using large sample theory) standard errors for a one-factor model.

Table 7.3 *Parameter estimates and standard errors in brackets and standardized loadings for the one-factor model, attitude to abortion*

Item	$\hat{\alpha}_{i0}$	s.e.	$\hat{\alpha}_{i1}$	s.e.	st$\hat{\alpha}_{i1}$	$\hat{\pi}_i(0)$
WomanDecide	−0.72	(0.33)	4.15	(0.85)	0.97	0.33
CoupleDecide	1.11	(0.35)	4.50	(0.81)	0.98	0.75
NotMarried	2.18	(0.61)	6.21	(1.54)	0.99	0.90
CannotAfford	1.15	(0.28)	3.49	(0.50)	0.96	0.76

The last column of the table gives the estimated probabilities that the median individual will respond positively to items 1-4. Item 1 stands out from the other items by being much less likely to be answered positively by the median individual. The loadings in the $\hat{\alpha}_{i1}$ column are all positive and very large, suggesting an underlying factor which is common to all items. In this context, one might identify this with a pro/anti-abortion attitude. It should be noted that the standard errors are all fairly large in relation to the differences in the estimates. This should caution us against placing undue weight on small inequalities among the loadings. In the present case, the broad conclusion we have drawn about a common factor seems unlikely to be sensitive to the effects of sampling variation. Taking the loadings at their face value, it appears that "the couple being unmarried" is the best discriminator between a pro- and an anti-abortion attitude and "inability to afford the baby" the worst discriminator.

The column headed st$\hat{\alpha}_{i1}$ requires some further explanation. In factor analysis, when the correlation matrix is analyzed, the factor loading α_{ij} is the correlation between the observed item x_i and the latent variable y_j. This was very convenient as an aid to interpretation. In the latent trait case, the loadings cannot be interpreted as correlation coefficients; indeed, as we have seen, the loadings are not bounded by 0 and 1 as a correlation would be. However, it is possible to transform the loadings so that they can be interpreted as correlation coefficients in exactly the same way as in factor analysis. This transformation arises naturally out of the alternative way of deriving latent trait models which we shall consider in Section 7.7. We shall defer consideration of this point to later, but here we merely observe that all the standardized loadings are close to one, indicating a close link between each item and the common factor.

7.4 Goodness-of-fit

The goodness-of-fit of the model can be checked in several different ways.

i) *Global goodness-of-fit test*
One way is to use a standard goodness-of-fit test to compare the observed and expected frequencies across the response patterns. Strictly, we compare observed frequencies and estimates of the expected frequencies under the model being tested — but conventionally, these estimates are referred to as "expected frequencies" when carrying out likelihood ratio or Pearson chi-squared goodness-of-fit tests as below. In fact, since we fit the models by choosing the parameter values so that these distributions are as close as possible, the minimum closeness would be an obvious measure to use for goodness-of-fit. A test based on such a measure is the log-likelihood-ratio test. The log-likelihood-ratio test statistic, G^2, is defined as:

$$G^2 = 2 \sum_{r=1}^{2^p} O(r) \log_e \frac{O(r)}{E(r)} \qquad (7.5)$$

where r represents a response pattern, and $O(r)$ and $E(r)$ represent the observed and expected frequencies, respectively, of response pattern r. An alternative is to use the Pearson chi-squared goodness-of-fit test statistic, X^2, given by:

$$X^2 = \sum_{r=1}^{2^p} \frac{(O(r) - E(r))^2}{E(r)}. \qquad (7.6)$$

If the model holds, both statistics are distributed approximately as χ^2 with degrees of freedom equal to the number of different response patterns minus the number of independent parameters minus one $(2^p - p(q+1) - 1)$. If the sample size n is much bigger than the total number of distinct responses given by 2^p, then the observed and expected frequencies will be reasonably large and the approximation on which the test is based will be valid. However, when the number of binary variables is large, many response patterns will have expected frequencies which are very small. It is usually recommended that all expected frequencies should be at least five for either test to be valid. If, for example, $p = 20$, there are $2^p = 1048576$ possible response patterns, and even with a sample size of several thousands there will be many expected frequencies which are exceedingly small. In those cases, the chi-squared test and the log-likelihood-ratio test will not follow a chi-squared distribution, and so from the practical point of view these tests cannot be used. The problem can be overcome to some extent by pooling response patterns with expected frequencies less than 5, but that might quickly lead to a situation where no degrees of freedom are left to perform the test. In such cases, we need another approach.

For the attitude to abortion data set, the log-likelihood-ratio statistic is $G^2 = 17.85$ and the chi-squared statistic is $X^2 = 15.09$ both on three degrees of freedom. (This is not the seven degrees of freedom one would have expected from the formula $2^p - 2p - 1$, because some grouping of categories has taken place.)

Both measures indicate a not very good fit (the 1% significance level for chi-squared with three degrees of freedom is 11.35). We should be cautious when we interpret those two statistics. The chi-squared test might be questionable in this case because there are several response patterns with expected frequencies less than 5 (see Table 7.6, column two). We could go on to fit a two-factor model, but first it is worth trying to diagnose the reason for the poor fit. The first step is obviously to look for large discrepancies between observed and expected score patterns. These are given in the first two columns of Table 7.6. There are no obviously large deviations except, perhaps, at the two extremes. In a sparse table with many more response patterns it would be much more difficult to judge this, and then other approaches are needed.

ii) *Goodness-of-fit for margins*

Rather than look at the whole set of response patterns, we can look at the two-way margins. That is, we can construct the 2×2 contingency tables obtained by taking the variables two at a time. We have already done this at the beginning of the chapter when we looked at the pairwise associations among variables. The reason for doing that was to bring out the parallel with factor analysis. The two-way tables provided the same sort of information for binary variables as the correlations do for factor analysis. The two-way margins are the cell frequencies in these two-way tables. We do not need all four cell frequencies; one of them will suffice. This is because once one cell frequency is given, the other three are determined by subtraction from the row and column totals which are regarded as fixed. Comparing the observed and expected two-way margins is therefore analogous to comparing the observed and expected correlations when judging the fit of a factor model. The comparison is made using what we call *chi-squared residuals*. These are the contributions to the chi-squared statistic for the 2×2 table which would arise from the cell. Thus if O is the observed frequency and E the expected frequency, then the residual is $(O - E)^2/E$. Tables 7.4 and 7.5 give the observed and expected frequencies for the two-way and for some of the three-way margins respectively for the attitude to abortion data when the one-factor model is fitted. The last column of the tables gives the chi-squared residual as a measure of the discrepancy between the observed and the predicted frequency. From Table 7.4, we see that 147 respondents responded negatively to items 1 and 2; the model predicted 143.74 responses for that cell giving a residual equal to 0.07. The same calculations are done for all the pairs of items. The residuals computed for each cell are not independent and therefore they cannot be summed to give an overall test distributed as chi-squared. A valid test is, however, provided in Bartholomew and Leung (2002) . As a rule of thumb, if we consider the residual in each cell as having a χ^2 distribution with one degree of freedom, then a value of the residual greater than 4 is indicative of poor fit at the 5% significance level. To be able to have a better idea of the discrepancies in the margins, given that the value 4 is only indicative, in the examples later in the chapter, we also report residuals greater than 3. A study of the individual margins provides information about where the model does not fit. For the abortion data, all the residuals are very small. On the evidence from the margins, we have no reason to reject the one-factor model. The overall significant result we obtained from the global goodness-of-fit tests cannot therefore be attributed to the relationships between the pairs and triplets of items.

iii) *Proportion of G^2 explained*

We have remarked at several points in the book that even an incomplete summary of multivariate data can be useful. The same is true of a multivariate model. Even though it may leave something unexplained, it may nevertheless capture some important and interesting features of the data. This is the case with the one-factor model which is serving as our example in this section. This raises the question of whether we can quantify the degree to which a simple model explains the associations between the binary variables. The same general idea proved useful in PCA and FA, where the proportion of variance explained served a similar purpose. Thus, we observed that the proportion of the total variance accounted for by a set of components might be used as a guide to whether the ys were an adequate summary. The same idea can be used here, but we now talk in terms of the proportion of the log-likelihood-ratio statistic for the independence model, which is explained by the model with q factors. The independence model would be appropriate if there were no associations between the binary variables $x_1, \ldots, x_p$. The log-likelihood-ratio statistic, G_0^2, for this model can be regarded as a measure of the associations between the xs. The log-likelihood-ratio statistic, G_q^2, for the model with q latent variables is a measure of the residual associations between the xs which have not been explained by the model.

The percentage of G^2 explained is given by

$$\%G^2 = \frac{G_0^2 - G_q^2}{G_0^2} \times 100$$

Table 7.4 *Chi-squared residuals for the second order margins for the one-factor model, attitude towards abortion*

Response	Item i	Item j	Observed frequency (O)	Expected frequency (E)	$O - E$	$(O - E)^2/E$
(0,0)	2	1	147	143.74	3.26	0.07
	3	1	131	133.17	−2.17	0.04
	3	2	117	119.69	−2.69	0.06
	4	1	129	133.68	−4.68	0.16
	4	2	114	116.09	−2.09	0.04
	4	3	116	111.79	4.21	0.16
(0,1)	2	1	7	11.30	−4.30	1.64
	3	1	7	5.94	1.06	0.19
	3	2	21	19.42	1.58	0.13
	4	1	16	11.99	4.01	1.34
	4	2	31	29.58	1.42	0.07
	4	3	29	33.88	−4.88	0.70
(1,0)	2	1	66	69.89	−3.89	0.22
	3	1	82	80.46	1.54	0.03
	3	2	37	35.35	1.65	0.08
	4	1	84	79.95	4.05	0.21
	4	2	40	38.95	1.05	0.03
	4	3	22	27.32	−5.32	1.04
(1,1)	2	1	159	154.07	4.93	0.16
	3	1	159	159.43	−0.43	0.00
	3	2	204	204.54	−0.54	0.00
	4	1	150	153.38	−3.38	0.07
	4	2	194	194.38	−0.38	0.00
	4	3	212	206.01	5.99	0.17

Table 7.5 *Chi-squared residuals for the third order margins for the one-factor model, response (1,1,1) to items (i, j, k), attitude towards abortion*

Item i	Item j	Item k	Observed frequency (O)	Expected frequency (E)	$O - E$	$(O - E)^2/E$
1	2	3	153	151.18	1.82	0.02
1	2	4	144	145.86	−1.86	0.02
1	3	4	147	150.15	−3.15	0.07
2	3	4	185	185.01	−0.01	0.00

and measures the extent to which the model with q latent variables explains the associations.

For the attitude to abortion data, the percentage of G^2 explained is 96.88%, indicating that the one latent variable model is a much better fit than the independence model or, in other words, there is 96.88% reduction in the log-likelihood-ratio statistic when the one-factor model was fitted.

The above three ways of checking model fit have been discussed in detail in the paper by Bartholomew and Tzamourani (1999).

iv) *Model selection methods*
Another approach, already mentioned in the connection with factor analysis for metrical variables, is based on the use of model selection criteria such as the Akaike information criterion or the Bayesian information criterion (see Sclove 1987). As noted there, this method lies beyond the scope of the present book.

7.5 Factor scores

Obtaining factor scores for the latent trait model is slightly more complicated than it was for PCA or FA. In PCA, the scores came "ready-made" as linear combinations of the manifest variables. In FA, the position was complicated by the fact that there was no unique value of each y associated with the set of xs. We therefore used a predicted value which turned out to be a linear combination of the xs for which the coefficients were calculated by the standard software. Following the same idea for the latent trait model, we would look for a suitable predictor of each y given the xs. Using regression ideas as before, this would suggest using the conditional mean value or conditional expectation:

$$E(y_j \mid x_1, \ldots, x_p) \qquad (j = 1, \ldots, q). \qquad (7.7)$$

Unfortunately, these means are not linear combinations of the xs, although they can easily be computed. However, it turns out that (for the logit link function) they are monotonic functions of what we shall call *component scores* which are given by:

$$X_j = \sum_{i=1}^{p} \alpha_{ij} x_i \qquad (j = 1, \ldots, q). \qquad (7.8)$$

In the one-factor case, both the regression function of equation (7.7) and the components give the same ranking to the individuals in the sample. These components are very simply calculated using the estimated weights obtained from fitting the model. For most practical purposes, it makes no difference whether we use the components or the conditional expectations.

For the logit link function, the component score, X_j, includes all the information in the data about the latent variables regardless of the assumption made about the distribution of y_j, whereas the posterior mean $E(y_j \mid x_1, \ldots, x_p)$ itself will vary according to whether we assume the distribution of y_j to be normal or some other distribution. This invariance property is a good reason for preferring the component score.

On the other hand, when a distribution is assumed for the ys it is possible to estimate not only the conditional means, $E(y_j \mid x_1, \ldots, x_p)$, but also the conditional standard deviations, $\sigma(y_j \mid x_1, \ldots, x_p)$ for $(j = 1, \ldots, q)$. The estimated standard deviations should be taken into account in judging the ranking of the response patterns when the conditional means are used.

Table 7.6 gives the estimated conditional means and component scores for all the response patterns for the attitude to abortion data. It also gives the expected frequency for each pattern. The sixth column gives the total score of the response pattern. As we can see, the estimated conditional mean and the component score give the same ranking to the individuals. In this particular example, the total score also gives a similar ranking to the individuals, though there are some ties. The reason for this is that all the four items have similar discriminating power. There is also a column headed $\hat{\sigma}(y \mid \mathbf{x})$. This is the estimated conditional standard deviation of the latent variable about its conditional mean. This tends to be larger at the extremes but is fairly constant over the middle range. In all cases it is quite large, indicating that the factor scores are subject to a good deal of uncertainty.

Table 7.6 *Factor scores listed in increasing order, attitude towards abortion*

Observed frequency	Expected frequency	$\hat{E}(y \mid \mathbf{x})$	$\hat{\sigma}(y \mid \mathbf{x})$	Component score (X_1)	Total score	Response pattern
103	100.0	−1.19	0.55	0.00	0	0000
13	16.6	−0.61	0.32	3.49	1	0001
1	1.7	−0.55	0.30	4.15	1	1000
9	9.1	−0.52	0.29	4.50	1	0100
10	12.3	−0.38	0.26	6.21	1	0010
0	1.3	−0.29	0.24	7.64	2	1001
6	7.4	−0.27	0.24	7.99	2	0101
3	1.0	−0.24	0.24	8.65	2	1100
21	14.8	−0.18	0.24	9.70	2	0011
0	2.0	−0.14	0.25	10.37	2	1010
7	12.3	−0.12	0.26	10.71	2	0110
3	1.9	−0.01	0.28	12.14	3	1101
6	6.2	0.14	0.32	13.86	3	1011
44	41.1	0.17	0.32	14.20	3	0111
12	7.2	0.24	0.34	14.87	3	1110
141	143.9	0.95	0.61	18.35	4	1111

7.6 Rotation

As with the factor analysis model, the solution is not unique when we fit more than one latent variable. An orthogonal rotation of the factors coupled with corresponding rotation of the estimated loadings $\hat{\alpha}_{ij}$ leaves the likelihood unchanged. We are therefore free to search for a rotation which is more readily interpretable. The cautionary remarks made in Chapter 6 apply with equal force here. In particular, rotation does not produce a new solution so much as express the original solution in a different way. The main use of rotation is to search for "simple structure". In principle, the same kind of rotations could be used for latent trait models as for factor analysis. However, the uncertainties of estimation increase rapidly with the number of factors. It is doubtful whether there is any value in trying to fit more than two factors with the sample sizes that are commonly available. In any case, we have concentrated in this book on solutions which are capable of being represented in up to two dimensions. Our treatment is therefore consistent with this general approach. For practical purposes, rotation can be carried out in two dimensions graphically, as in Chapter 6.

7.7 Underlying variable approach

In this section we will discuss the alternative approach for constructing and fitting a latent trait model to binary items. This approach is called the underlying variable (UV) approach. As we explained in Section 7.2, the UV approach is closer in spirit to factor analysis.

In the UV approach, the observed binary variables are assumed to be realizations of fictitious continuous *underlying variables*. Those underlying variables are unobserved but they should not be confused with the latent variables. They might be better described as *incompletely observed variables*, because all we observe is whether or not they exceed some threshold.

For each binary variable x_i, it is assumed that there is an *incompletely observed* continuous variable x_i^* which is normally distributed with mean μ_i and variance σ_i^2.

The connection between x_i and x_i^* is as follows: when the underlying variable x_i^* takes values below a threshold value τ_i, the binary item x_i takes the value 1, otherwise x_i takes the value 0. The parameters τ_i are called threshold parameters. Since no information is available about x_i^*, beyond that given above, its mean μ_i and variance σ_i^2 cannot be determined and are therefore, by convention, set to zero and one, respectively.

The essence of the method is to treat the x_i^*s as if they had been generated by the classical factor analysis model. That is, we suppose that:

$$x_i^* = \alpha_{i0}^* + \alpha_{i1}^* y_1 + \alpha_{i2}^* y_3 + \cdots + \alpha_{iq}^* y_q + e_i \qquad (i = 1, \ldots, p), \qquad (7.9)$$

where α_{ij}^* $(j = 1, \ldots, q)$ are the factor loadings, the y_js are the latent variables, and e_i $(i = 1, \ldots, p)$ are the residuals. In factor analysis, the x_i^*s are observable variables, whereas here they are underlying, unobservable variables.

All that we need to fit a factor model is the matrix of correlations. The correlation can be estimated from each pair of the binary x_is and hence software such as LISREL, Mplus, and EQS can be used to fit the factor model. Correlations estimated in this way are called *tetrachoric* correlations.

There are a number of subtle differences between the fitting of a factor model to tetrachoric correlations and fitting it to product moment correlations. The UV approach does not specify a model for the complete p-dimensional response pattern as is the case in the response theory approach. The thresholds are estimated from the univariate marginal distribution of the underlying variable, x_i^*, and the correlations from the bivariate marginal distributions of the x_i^*s for given thresholds. This amounts to saying that the method uses less of the information in the data. The UV approach does make the assumption of conditional independence through the independence of the residual terms, e_i, and it also assumes that the univariate and bivariate distributions of the underlying variables are normal.

The results of carrying out a factor analysis on tetrachoric correlations are very similar to those obtained using the logit latent variable model. This is no accident, because it can be shown that the two types of model are equivalent for binary data. A mathematical proof of this equivalence will be found in Bartholomew and Knott (1999) , p.87-88. In the present instance, it means that if we had chosen the underlying distribution of the manifest variables to be standard logistic rather than normal, the resulting models would have been identical in the two cases. The logit model and the UV normit model give such similar results because the normal and logistic distributions are so similar in shape. There is an exact equivalence between the parameter estimates for the normit UV and the normit IRF model given by:

$$\alpha_{i0} = \frac{(\tau_i - \alpha_{i0}^*)}{\sigma_i}$$

and

$$\alpha_{ij} = -\frac{\alpha_{ij}^*}{\sigma_i}.$$

Furthermore, we can standardize the factor loadings α_{ij} to represent correlation coefficients between the latent variables and the underlying continuous variables x_i^*.

The standardized αs are given by:

$$\mathrm{st}\alpha_{ij} = \frac{\alpha_{ij}}{\sqrt{\sum_{j=1}^{q} \alpha_{ij}^2 + 1}}. \tag{7.10}$$

This is the standardisation we referred to in Section 7.3 and which was given in Table 7.3.

7.8 Example: sexual attitudes

In order to illustrate the full range of analyses, including the fitting of two factors, we shall take an example extracted from the 1990 British Social Attitudes Survey (Brook, Taylor, and Prior 1991) . It concerns contemporary sexual attitudes. The questions addressed to 1077 individuals were as follows.

1. Should divorce be easier?
2. Do you support the law against sexual discrimination?
3. View on pre-marital sex: not at all wrong...always wrong.
4. View on extra-marital sex: not at all wrong...always wrong.
5. View on sexual relationship between individuals of the same sex: not at all wrong...always wrong.
6. Should gays teach in school?
7. Should gays teach in higher education?
8. Should gays hold public positions?
9. Should a female homosexual couple be allowed to adopt children?
10. Should a male homosexual couple be allowed to adopt children?

For those items yielding a binary response (1,2,6,7,8,9,10), a positive response was coded as 1 and a negative response as 0. For items 3, 4, and 5 there were five categories: "always wrong", "mostly wrong", "sometimes wrong", "rarely wrong" and "not at all wrong". Responses "sometimes wrong", "rarely wrong", and "not at all wrong" were coded as 1 and responses "always wrong" and "mostly wrong" as 0. With ten variables, there are $2^{10} = 1024$ possible response patterns. Not all of these occur, but with a sample size of 1077 the data matrix takes up a good deal of space. The full data set is given on the Web site, but the cases with frequencies greater than ten are listed in Table 7.7 in decreasing order of observed frequency as an illustration.

Table 7.8 gives the proportions giving positive and negative responses to each item.

Since we come to the data with no preconceived ideas about what the latent variables might be, we begin by fitting a one-factor model to the ten items. The parameter estimates are listed in Table 7.9. Items 6, 7, and 8 have large discrimination coefficients, indicating that the characteristic curves of those items are very steep. From the $st\hat{\alpha}_{i1}$ column, we see that item 1 has the weakest relationship with the latent variable, followed by items 2 and 4. The rest of the items show strong relationships with the latent variable y.

We first investigate the goodness-of-fit of the one-factor model using the methods described in Section 7.4. They all suggest that the one-factor model is not a satisfactory fit to the data. The overall goodness-of-fit measures suggested a very bad fit ($G^2 = 427.39, X^2 = 354.30$ on 32 degrees of freedom). There were also large discrepancies between the observed and expected frequencies for many pairs and triplets of items. Table 7.10 gives all the pairs and the (1,1,1) triplets of items where the chi-squared residuals were greater than 3.

The percentage of G^2 explained is 77.03%, which shows that the model goes a long way in explaining the associations, but taken with the very poor

Table 7.7 *Response frequencies, sexual attitudes data*

Response patterns	Frequency	Response patterns	Frequency
0110000000	117	1110000000	17
0110111100	95	0111000000	15
0100000000	93	0111011100	15
0110011100	90	0010011100	14
0110111111	40	0110111110	14
0010000000	35	0110011110	13
0100011100	32	1110011100	13
0000000000	29	1110111111	12
0110000100	27	0110011000	11
0110001100	21	0110100000	11
0111111100	19	0010000100	11
0100000100	18	0000011100	10
0111111111	18	Other patterns	287

Table 7.8 *Proportions giving positive and negative responses to observed items, sexual attitudes data*

Item	Response 1	Response 0
1	0.13	0.87
2	0.83	0.18
3	0.77	0.23
4	0.13	0.87
5	0.29	0.71
6	0.48	0.53
7	0.55	0.45
8	0.59	0.41
9	0.19	0.81
10	0.11	0.89

fit indicated by the other tests it is clearly desirable to continue by fitting a second latent variable.

The two-factor model is a considerable improvement. The percentage of G^2 explained increased from 77.03 to 86.8%. The log-likelihood-ratio statistic and the chi-squared statistic still indicate a poor fit ($G^2 = 268.50, X^2 = 199.07$, each on 24 degrees of freedom). However, we need to look at the fit on the margins before making a final judgement.

Comparing the results from the one-factor solution given in Table 7.10, we find that the two-factor solution is a great improvement for predicting the

Table 7.9 *Estimated difficulty and discrimination parameters with standard errors in brackets and standardized loadings for the one-factor model, sexual attitudes data*

Items	$\hat{\alpha}_{i0}$	s.e.	$\hat{\alpha}_{i1}$	s.e.	st$\hat{\alpha}_{i1}$	$\hat{\pi}_i(0)$
1	-1.93	(0.09)	0.11	(0.10)	0.11	0.13
2	1.65	(0.10)	0.53	(0.11)	0.47	0.84
3	1.46	(0.10)	1.00	(0.11)	0.71	0.81
4	-2.01	(0.11)	0.60	(0.10)	0.52	0.12
5	-1.29	(0.11)	1.79	(0.16)	0.87	0.22
6	-0.12	(0.45)	10.08	(1.63)	1.00	0.47
7	1.99	(0.84)	10.05	(3.39)	1.00	0.88
8	1.05	(0.17)	3.52	(0.30)	0.96	0.74
9	-2.06	(0.14)	1.64	(0.18)	0.85	0.11
10	-3.72	(0.27)	2.44	(0.25)	0.93	0.02

observed two- and three-way margins. The fit was found to be poor (with residuals greater than 3) for the margins given in Table 7.11.

Although the fit is still somewhat questionable, the large percentage of G^2 explained encourages us to attempt an interpretation of the two-factor model.

Table 7.12 gives the maximum likelihood estimates together with their asymptotic (estimated using large sample theory) standard errors and the standardized parameters for the factor loadings. The last column shows very striking differences in the response of the median individual to the various questions. The last two items on adoption by homosexual couples show virtually no support for the propositions. There are also small probabilities of responding positively to items 1, 4, and 5. The marginal observed proportions given in Table 7.8 give a similar picture but they relate to views in the whole population rather than to the median individual. As an aid to interpretation, the standardized factor loadings are plotted in Figure 7.3

We see that items 2, 6, 7, and 8 have high loadings on the first factor and low loadings on the second factor. Items 3, 4 , 9, and 10 have high loadings on the second factor and low on the first factor. Item 5 lies somewhere in between. The interpretation is not entirely clear, but we note that the items in the first group are concerned with public matters whereas items 2 and 6, at least, are concerned with private behaviour. However, the inclusion of items 9 and 10 does not fit with this interpretation. We might hope that the plot of the loadings would suggest a rotation that would help the interpretation. From Figure 7.3, we see that there is no obvious orthogonal rotation that produces a simpler pattern than the one revealed from the original factor solution.

The failure to get two clear-cut factors coupled with the poor fit of the model overall suggests that the analysis should be taken further. The obvious thing would be to try a three-factor model or to re-analyze the data omitting the last two items, which seem to differ in some fundamental way from the

Table 7.10 *Chi-squared residuals greater than 3 for all the second order and (1,1,1) third order margins for the one-factor model, sexual attitudes data*

Response	Items	O	E	$O-E$	$(O-E)^2/E$
(0,0)	5, 3	237	208.20	28.80	3.98
	10, 9	875	814.79	60.23	4.45
(0,1)	3, 1	19	29.08	−10.08	3.49
	10, 9	88	144.77	−56.77	22.26
(1,0)	4, 3	4	22.46	−18.46	15.17
	5, 2	23	37.85	−14.85	5.83
	5, 3	14	36.52	−22.52	13.89
	9, 6	46	25.95	20.05	15.50
	9, 7	36	17.88	18.12	18.35
	9, 8	29	19.87	9.13	4.20
	10, 5	23	34.17	−11.17	3.65
	10, 6	15	4.13	10.87	28.66
	10, 7	12	2.50	9.50	36.16
	10, 8	11	3.73	7.27	14.19
	10, 9	2	54.36	−52.36	50.44
(1,1)	4, 1	29	18.88	10.12	5.42
	9, 6	154	181.92	−27.92	4.29
	9, 7	164	189.98	−25.98	3.55
	10, 9	112	63.09	48.91	37.91
(1,1,1)	1, 2, 6	50	64.53	−14.53	3.27
	1, 3, 4	29	16.10	12.90	10.34
	1, 4, 8	22	14.94	7.06	3.33
	1, 4, 10	8	4.18	3.82	3.49
	1, 5, 10	20	12.21	7.79	4.97
	1, 9, 10	21	9.33	11.67	14.60
	2, 3, 4	122	104.17	17.83	3.05
	2, 6, 9	137	164.02	−27.02	4.45
	2, 7, 9	147	170.74	−23.74	3.30
	2, 9, 10	99	58.25	40.75	28.51
	3, 9, 10	106	60.00	46.00	35.26
	4, 9, 10	33	17.31	15.67	14.21
	5, 9, 10	89	50.37	38.63	29.63
	6, 7, 9	153	180.82	−27.82	4.28
	6, 8, 9	151	176.04	−25.04	3.56
	6, 9, 10	97	62.74	34.26	18.71
	7, 9, 10	100	62.92	37.08	21.85
	8, 9, 10	101	62.55	38.45	23.64

Table 7.11 *Chi-squared residuals greater than 3 for the second and (1,1,1) third order margins for the two-factor model, sexual attitudes data*

Response	Items	O	E	$O - E$	$(O - E)^2/E$
(0,0)	7, 6	477	436.9	40.01	3.66
	9, 7	451	413.58	37.42	3.38
	7, 7	487	448.17	38.83	3.37
	8, 7	382	349.33	32.67	3.38
	10, 7	475	436.50	38.49	3.39
(1,0)	4, 3	4	17.65	−13.65	10.55
	5, 2	23	38.38	−15.38	6.16
	5, 3	14	28.06	−14.06	7.04
	10, 3	6	2.62	3.38	4.35
(1,1,1)	1, 3, 4	29	19.51	9.49	4.62

Table 7.12 *Estimated difficulty and discrimination parameters with standard errors in brackets and standardized loadings for the two-factor model, sexual attitudes data*

Items	$\hat{\alpha}_{i0}$	s.e.	$\hat{\alpha}_{i1}$	s.e.	$\hat{\alpha}_{i2}$	s.e.	st$\hat{\alpha}_{i1}$	st$\hat{\alpha}_{i2}$	$\hat{\pi}_i(0)$
1	−2.01	(0.11)	−0.25	(0.14)	0.38	(0.13)	−0.22	0.35	0.12
2	1.67	(0.09)	0.51	(0.12)	0.22	(0.12)	0.44	0.19	0.84
3	1.64	(0.12)	0.40	(0.13)	1.30	(0.16)	0.24	0.77	0.84
4	−2.10	(0.12)	0.11	(0.12)	0.79	(0.14)	0.09	0.62	0.11
5	−1.40	(0.13)	1.12	(0.14)	1.65	(0.17)	0.50	0.74	0.20
6	−0.05	(0.34)	8.12	(1.65)	4.41	(0.88)	0.87	0.48	0.49
7	2.46	(1.46)	10.26	(5.48)	6.22	(2.78)	0.85	0.52	0.92
8	1.06	(0.15)	2.79	(0.26)	1.83	(0.21)	0.80	0.53	0.74
9	−4.14	(0.71)	0.11	(0.23)	4.86	(1.20)	0.02	0.98	0.02
10	−14.82	(202.11)	0.54	(0.77)	10.22	(123.60)	0.05	0.99	0.00

other items. A third possibility, to which we shall return in Chapter 9, is to consider a different kind of model; namely, a latent class model.

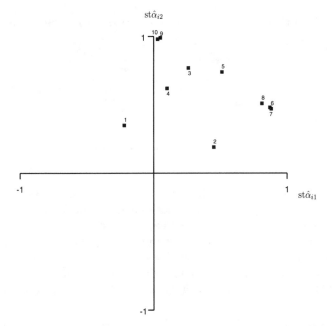

Figure 7.3 *Plots of standardized loadings, sexual attitudes data*

7.9 Further examples and suggestions for further work

The Law School Admission Test (LSAT), Section VI

The LSAT example is part of an educational test data set given in Bock and Lieberman (1970) . The LSAT is a classical example in educational testing for measuring ability traits. This is a test that was designed to measure a *single* latent ability scale. The test as given in Bock and Lieberman (1970) consisted of five items taken by 1000 individuals. The main interest is whether the attempt to construct items which are indicators solely of this ability has been successful and, if so, what do the parameter estimates tell us about the items. From Table 7.13, you will see that 92% of the students answered item 1 correctly but only 55% answered item 3 correctly. That makes item 3 the most "difficult" among the five items. The full data set is given in Table 7.15. To investigate whether the five items form a *unidimensional* scale, you need to test whether the one-factor model is a good fit to the five items. The overall goodness-of-fit measures show that the one-factor model is a very good fit to the data ($G^2 = 15.30$ and $X^2 = 11.66$ on 13 degrees of freedom). In other words, the associations among the five items can be explained by a single latent variable that in this example is an ability which the test is designed to measure. Since the one-factor model is not rejected by the overall goodness-of-fit test, there is no need to check the fit on the two- and three-way margins. G^2 and X^2 measure how well the model predicts the whole response pattern. The

Table 7.13 *Proportions of positive and negative responses for observed items, LSAT data*

Item	Response 1	Response 0
1	0.92	0.08
2	0.71	0.29
3	0.55	0.45
4	0.76	0.24
5	0.87	0.13

first two columns of Table 7.15 show small discrepancies between the observed frequencies and the expected frequencies under the one-factor model.

Table 7.14 gives the parameter estimates for the one-factor solution. The last column of the table, $\hat{\pi}_i(0)$, gives the probability that the median individual will respond correctly to any of those five items. The five items have different difficulty levels. However the median individual has quite a high chance of getting the items correct indicating that, overall, the items are quite easy.

Table 7.14 *Estimated difficulty and discrimination parameters with standard errors in brackets and standardized loadings for the one-factor model, LSAT data*

item	$\hat{\alpha}_{i0}$	s.e.	$\hat{\alpha}_{i1}$	s.e.	st$\hat{\alpha}_{i1}$	$\hat{\pi}_i(0)$
1	2.77	(0.20)	0.83	(0.25)	0.64	0.94
2	0.99	(0.09)	0.72	(0.19)	0.59	0.73
3	0.25	(0.08)	0.89	(0.23)	0.67	0.56
4	1.28	(0.10)	0.69	(0.19)	0.57	0.78
5	2.05	(0.13)	0.66	(0.20)	0.55	0.89

The factor loadings $\hat{\alpha}_{i1}$ are all positive and of similar magnitude with similar standard errors. The same is true for the standardized loadings st$\hat{\alpha}_{i1}$. That implies that all five items have similar discriminating power and so a similar weight is applied to each response. In that case, the component score that is used to scale individuals on the latent dimension should give results that are close (similar ranking) to the scores obtained when the total score is used (see columns five and six of Table 7.15). You may analyze the five items using a latent class model that is discussed in Chapter 9 and compare the ranking of the individuals obtained from the latent trait model with the allocation of individuals into two distinct classes.

Table 7.15 *Factor scores in increasing order, LSAT data*

Observed frequency	Expected frequency	$\hat{E}(y \mid \mathbf{x})$	$\hat{\sigma}(y \mid \mathbf{x})$	Component score (X_1)	Total score	Response pattern
3	2.3	−1.90	0.80	0.00	0	00000
6	5.9	−1.47	0.80	0.66	1	00001
2	2.6	−1.45	0.80	0.69	1	00010
1	1.8	−1.43	0.80	0.72	1	01000
10	9.5	−1.37	0.80	0.83	1	10000
1	0.7	−1.32	0.80	0.89	1	00100
11	8.9	−1.03	0.81	1.35	2	00011
8	6.4	−1.01	0.81	1.38	2	01001
29	34.6	−0.94	0.81	1.48	2	10001
14	15.6	−0.92	0.81	1.51	2	10010
1	2.6	−0.90	0.81	1.55	2	00101
16	11.3	−0.90	0.81	1.55	2	11000
3	1.2	−0.88	0.81	1.58	2	00110
3	4.7	−0.79	0.81	1.72	2	10100
16	13.6	−0.55	0.82	2.07	3	01011
81	76.6	−0.48	0.82	2.17	3	10011
56	56.1	−0.46	0.82	2.21	3	11001
4	6.0	−0.44	0.82	2.24	3	00111
21	25.7	−0.44	0.82	2.24	3	11010
3	4.4	−0.42	0.82	2.27	3	01101
2	2.0	−0.40	0.82	2.30	3	01110
28	25.0	−0.35	0.82	2.37	3	10101
15	11.5	−0.33	0.82	2.40	3	10110
11	8.4	−0.30	0.82	2.44	3	11100
173	173.3	0.01	0.83	2.89	4	11011
15	13.9	0.05	0.84	2.96	4	01111
80	83.5	0.13	0.84	3.06	4	10111
61	62.5	0.15	0.84	3.10	4	11101
28	29.1	0.17	0.84	3.13	4	11110
298	296.7	0.65	0.86	3.78	5	11111

Workplace industrial relations data

This example is taken from a section of the 1990 Workplace Industrial Relations Survey (WIRS) dealing with management/worker consultation in firms. The WIRS surveys can be found at http://www.niesr.ac.uk/niesr/wers98/. A subset of the data is used here that consists of 1005 firms and concerns non-manual workers. The questions asked are given below:

Please consider the most recent change involving the introduction of new plant, machinery and equipment. Were discussions or consultations of any of the type on this card held either about the introduction of the change or about the way it was to be implemented?

1. Informal discussion with individual workers.
2. Meetings with groups of workers.
3. Discussions in established joint consultative committee.
4. Discussions in specially constituted committee to consider the change.
5. Discussions with union representatives at the establishment.
6. Discussions with paid union officials from outside.

All six items measure the amount of consultation that takes place in firms at different levels of the firm structure. Items 1 to 6 cover a range of informal to formal types of consultation. Those firms which place a high value on consultation might be expected to use all or most consultation practices. The six items are analyzed here using the latent trait model. We should mention that the items discussed here were not initially constructed to form a scale as is the case in the LSAT example and in most educational data. Therefore, our analysis is completely exploratory. The full data set is given on the Web site. The proportions giving positive and negative responses to each item are given in Table 7.16. The most common type of consultation among the 1005 firms is the established joint consultative committee. The one-factor model

Table 7.16 *Proportions giving positive and negative responses to observed items, WIRS data*

Item	Response 1	Response 0
1	0.37	0.63
2	0.58	0.42
3	0.28	0.72
4	0.24	0.76
5	0.36	0.64
6	0.15	0.85

gives $G^2 = 269.4$ and $X^2 = 264.2$ on 32 degrees of freedom. Both goodness-of-fit measures indicate that the one-factor model is a poor fit to the data. Table 7.17 gives chi-squared residuals greater than 3 for the second and third-way margins. The largest discrepancies are found between items 1 and 2. As a result, the model fails to explain the associations among the six items, judging by the overall goodness-of-fit measures, and it also fails to explain the pairwise associations.

You should continue the analysis by fitting one more latent variable that might account for the big discrepancies between the observed and expected frequencies. The percentage of G^2 explained increases from 49.35% for the one-factor model to 74.58% for the two-factor model. Clearly, the second latent variable contributes substantially in explaining the associations among the six items. However, the fit of the two-factor model is still poor if we look at the $G^2 = 146.4$ and $X^2 = 131.5$ on 24 degrees of freedom. However, the

Table 7.17 *Chi-squared residuals greater than 3 for the second and (1,1,1) third order margins for the one-factor model, WIRS data*

Response	Items	O	E	$O - E$	$(O - E)^2/E$
(0,0)	2, 1	186	265.75	−79.75	23.93
(0,1)	2, 1	233	153.23	79.77	41.52
(1,0)	2, 1	444	364.25	79.75	17.46
	4, 1	172	145.48	26.52	4.84
	4, 2	61	87.00	−26.00	7.77
(1,1)	2, 1	142	221.77	−79.77	28.69
	4, 1	69	95.65	−26.65	7.43
	4, 2	180	154.13	25.87	4.34
(1,1,1)	1, 2, 3	37	75.79	−38.79	19.85
	1 ,2, 4	23	61.75	−38.75	24.32
	1, 2, 5	53	94.85	−41.85	18.46
	1, 2, 6	26	40.32	−14.32	5.08
	1, 3, 4	30	45.69	−15.69	5.39
	1, 4, 5	35	55.73	−20.73	7.71
	2, 3, 4	93	75.03	17.97	4.31
	2, 4, 5	108	91.39	16.61	3.02

residuals for the two-way margins are all close to zero. The second latent variable accounts for the pairwise associations but the fit is still not satisfactory on the three-way margins. Table 7.18 gives the residuals greater than 3 for the $(1, 1, 1)$ three-way margins. Item 1 appears in all the triplets that show a bad fit. This is the least formal item, which is also vaguely worded and might be interpreted differently by different respondents.

Table 7.18 *Chi-squared residuals greater than 3 for the third order margins for the two-factor model, response (1,1,1) to items (i, j, k), WIRS data*

Item i	Item j	Item k	O	E	$O - E$	$(O - E)^2/E$
1	2	3	37	60.53	−23.53	9.15
1	2	4	23	40.65	−17.65	7.66
1	2	5	53	73.99	−20.99	5.95
1	3	6	31	42.54	−11.54	3.13
1	5	6	36	49.84	−13.84	3.84

Although the model is not good in predicting the three-way margins, it is worth looking at the parameter estimates of the two-factor latent trait model given in Table 7.19. All the loadings ($\hat{\alpha}_{i1}$) of the first factor except that for item 1 (the least formal item) are positive and large indicating a "general" factor relating to amount of consultation which takes place.

Table 7.19 *Estimated difficulty and discrimination parameters with standard errors in brackets and standardized loadings for the two-factor model, WIRS data*

Items	$\hat{\alpha}_{i0}$	s.e.	$\hat{\alpha}_{i1}$	s.e.	$\hat{\alpha}_{i2}$	s.e.	st$\hat{\alpha}_{i1}$	st$\hat{\alpha}_{i2}$	$\hat{\pi}_i(0)$
1	−0.93	(0.31)	−0.97	(0.48)	2.13	(0.96)	−0.38	0.84	0.28
2	0.54	(0.15)	1.51	(0.47)	−0.96	(0.36)	0.74	−0.47	0.63
3	−1.40	(0.14)	1.31	(0.18)	1.11	(0.18)	0.66	0.56	0.20
4	−1.47	(0.11)	1.22	(0.15)	0.12	(0.11)	0.77	0.08	0.19
5	−0.97	(0.14)	1.58	(0.24)	1.24	(0.21)	0.70	0.55	0.27
6	−2.39	(0.20)	1.05	(0.16)	1.06	(0.21)	0.59	0.59	0.08

The analysis may be repeated with item 1 omitted. The items used in the analysis are item 2 to item 6 and those names are used here. The one-factor model gives $G^2 = 50.50$ and $X^2 = 46.29$ on 17 degrees of freedom. The one-factor model is rejected. The fit of the two-way margins is very good except for two pairs, and there is only one large chi-squared residual in the (1,1,1) three-way margins. These residuals are given in Table 7.20 and all include item 2 which is the second least formal item after item 1 (which is omitted from the current analysis). The fit is improved when the two-factor model is fitted giving a $G^2 = 30.16$ and $X^2 = 27.53$ on 13 degrees of freedom. Those statistics still reject the two-factor model. However, the fit on the two-way margins is excellent and the $(1,1,1)$ three-way margins have no residual greater than 0.89. Further analysis of this data set can be found in Bartholomew (1998).

Table 7.20 *Chi-squared residuals greater than 3 for the second and the (1,1,1) third order margins for the one-factor model, WIRS data, item 1 omitted*

Response	Items	O	E	$O - E$	$(O - E)^2/E$
(1,0)	3, 1	61	84.94	−23.94	6.75
(1,1)	4, 2	180	156.28	23.72	3.60
(1,1,1)	(1, 2, 3)	93	77.09	15.91	3.28

The parameter estimates of the one-factor model given in Table 7.21 indicate a clear general factor corresponding to the amount of consultation that takes place. Note that item 2 has the smallest factor loading, while items 3 to 6 have similar factor loadings. It is quite apparent that items 3 to 6 can be considered separately to construct a scale measuring the amount of formal consultation which takes place. Fitting the one-factor model to items 3 to 6 gives $G^2 = 16.6$ and $X^2 = 14.5$ on seven degrees of freedom. All residuals for the two- and three-way margins are smaller than 1.0.

Table 7.21 *Estimated difficulty and discrimination parameters with standard errors in brackets and standardized loadings for the one-factor model with item 1 omitted, WIRS data*

item	$\hat{\alpha}_{i0}$	s.e.	$\hat{\alpha}_{i1}$	s.e.	st$\hat{\alpha}_{i1}$	$\hat{\pi}_i(0)$
2	0.35	(0.07)	0.42	(0.10)	0.39	0.59
3	−1.38	(0.14)	1.69	(0.23)	0.86	0.20
4	−1.40	(0.10)	1.05	(0.14)	0.72	0.20
5	−0.95	(0.13)	1.97	(0.31)	0.89	0.28
6	−2.29	(0.16)	1.34	(0.18)	0.80	0.09

Women's mobility

These data are from the Bangladesh Fertility Survey of 1989 (Huq and Cleland 1990) . The rural subsample of 8445 women is analyzed here. The questionnaire contains a number of items believed to measure different dimensions of women's status. The particular dimension that we shall focus on here is women's mobility or social freedom. Women were asked whether they could engage in the following activities alone (1=yes, 0=no).

1. Go to any part of the village/town/city.
2. Go outside the village/town/city.
3. Talk to a man you do not know.
4. Go to a cinema/cultural show.
5. Go shopping.
6. Go to a cooperative/mothers' club/other club.
7. Attend a political meeting.
8. Go to a health centre/hospital.

First, the one-factor model was fitted to the eight items to investigate whether the variables are all indicators of the same type of women's mobility in society. The one-factor model gives a G^2 equal to 364.5 on 39 degrees of freedom indicating a bad fit. Table 7.22 shows the chi-squared residuals greater than 3 for the two-way margins and the (1,1,1) three-way margins of the one-factor model.

The two-factor model is still rejected based on a G^2 equal to 263.41 on 33 degrees of freedom. The percentage of G^2 explained increases only slightly from 94.98% to 96.92%. However, although the contribution of the second factor is small, the fit on the two-way margins and the (1,1,1) three-way margins is generally very good; the margins for which the fit is poor are shown in Table 7.23

Table 7.22 *Chi-squared residuals greater than 3 for the second and the (1,1,1) third order margins for the one-factor model, women's mobility data*

Response	Items	O	E	$O - E$	$(O - E)^2/E$
(0,1)	3, 2	187	229.19	-42.19	7.76
	6, 2	1986	1899.91	86.09	3.90
	7, 6	532	596.04	-64.04	6.88
	8, 5	194	245.15	-51.15	10.67
	8, 7	108	134.51	-26.51	5.22
(1,0)	2, 1	52	117.29	-65.29	36.35
	5, 1	13	3.02	9.99	32.92
	5, 2	98	77.74	20.25	5.28
	5, 3	20	12.40	7.60	4.66
	5, 4	19	28.75	-9.75	3.31
	6, 2	274	196.34	77.66	30.71
	6, 3	44	32.03	11.97	4.47
	7, 1	6	1.13	4.87	20.97
	7, 2	62	36.82	25.18	17.21
	7, 4	17	8.75	8.25	7.78
	7, 6	41	93.69	-52.69	29.63
	8, 1	28	7.15	20.85	60.83
	8, 3	38	22.74	15.26	10.24
	8, 4	88	67.82	20.18	6.01
	8, 5	340	391.82	-51.82	6.85
(1,1)	6, 2	665	756.15	-91.15	10.99
	7, 6	407	356.45	50.55	7.17
	8, 5	392	348.29	43.71	5.48
(1,1,1)	1, 2, 3	2433	2338.67	94.33	3.80
	1, 2, 6	659	751.02	-92.02	11.27
	1, 5, 8	392	347.45	44.55	5.71
	1, 6, 7	403	355.75	47.25	6.27
	2, 3, 6	653	736.66	-83.66	9.50
	2, 4, 6	637	704.12	-67.12	6.40
	3, 5, 8	389	343.72	45.28	5.96
	3, 6, 7	402	352.32	49.68	7.01
	4, 5, 8	386	341.75	44.25	5.73
	4, 6, 7	396	351.63	44.37	5.60
	5, 6, 7	304	271.48	32.52	3.89
	5, 6, 8	326	279.56	46.44	7.72
	5, 7, 8	276	246.59	29.41	3.51
	6, 7, 8	318	267.09	50.91	9.70

Table 7.23 *Chi-squared residuals greater than 3 for the second and the (1,1,1) third order margins for the two-factor model, women's mobility data*

Response	Items	O	E	$O - E$	$(O - E)^2/E$
(0,1)	8, 5	194	239.58	-45.58	8.67
	8, 7	108	137.09	-29.09	6.17
(1,0)	4, 3	226	253.70	-27.70	3.02
	5, 1	13	7.12	5.88	4.86
	5, 4	19	33.25	-14.25	6.10
	6, 1	15	30.37	-15.37	7.78
	7, 2	62	78.03	-16.03	3.29
	7, 3	8	13.51	-5.51	2.25
	7, 6	41	67.28	-26.28	10.26
	8, 1	28	14.42	13.58	12.78
	8, 2	144	166.51	-22.51	3.04
	8, 4	88	71.84	16.16	3.64
	8, 5	340	388.56	-48.56	6.07
(1,1)	8, 5	392	355.73	36.27	3.70
(1,1,1)	1, 5, 8	392	353.37	38.63	4.22
	2, 5, 8	351	316.27	34.73	3.81
	3, 5, 8	389	348.32	40.68	4.75
	4, 5, 8	386	347.28	38.72	4.32
	5, 7, 8	276	245.75	30.25	3.72
	6, 7, 8	318	287.55	30.45	3.23

The parameter estimates for the two-factor model are given in Table 7.24. The eight items are positively correlated with both factors. However, as we can see from the standardized loadings $st\hat{\alpha}_{i1}$ and $st\hat{\alpha}_{i2}$, items 1 to 4 load heavily on the first factor where items 4 to 8 load heavily on the second factor. The loading for item 7 should be interpreted with caution due to its extremely large standard error. The two factors can be interpreted as measuring different dimensions of women's status. Items 5 to 8, and to some extent item 4, indicate a relatively high level of participation in public life; engaging in any of these activities would suggest a high degree of social freedom for a woman in rural Bangladesh. In contrast, items 1 to 3 are less specific indicating a degree of freedom but not necessarily in the public life sphere. The $\hat{\pi}_i(\mathbf{0})$ values show clearly that a woman who is in the middle of both factors has close to zero chances of responding positively to items 5 to 8. You should compare the results obtained here with those obtained in Chapter 9 where a latent class model is fitted.

Table 7.24 *Estimated difficulty and discrimination parameters with standard errors in brackets and standardized factor loadings for the two-factor model, women's mobility data*

Items	$\hat{\alpha}_{i0}$	s.e.	$\hat{\alpha}_{i1}$	s.e.	$\hat{\alpha}_{i2}$	s.e.	st$\hat{\alpha}_{i1}$	st$\hat{\alpha}_{i2}$	$\hat{\pi}_i(\mathbf{0})$
1	2.66	(0.18)	2.46	(0.28)	0.98	(0.17)	0.87	0.34	0.94
2	−1.58	(0.09)	2.48	(0.21)	1.32	(0.15)	0.83	0.44	0.17
3	1.56	(0.05)	1.25	(0.08)	0.86	(0.10)	0.69	0.47	0.83
4	−1.17	(0.06)	1.97	(0.16)	2.26	(0.17)	0.62	0.72	0.24
5	−6.58	(0.30)	1.98	(0.23)	3.57	(0.22)	0.47	0.85	0.00
6	−5.11	(0.27)	1.32	(0.23)	3.60	(0.24)	0.33	0.91	0.01
7	−17.24	(94.82)	2.20	(0.43)	10.01	(58.02)	0.21	0.97	0.00
8	−4.94	(0.17)	1.51	(0.17)	2.80	(0.15)	0.45	0.84	0.01

7.10 Software

The software GENLAT (Moustaki 2001) for implementing the item response function method is available on the Web site associated with the book. An important feature of the software is that it also produces estimated asymptotic standard errors for the estimates. These are based on asymptotic theory (large samples) and are only approximations but they often serve to add a note of caution to the interpretation, especially when the sample size is small. The program provides the goodness-of-fit measures and scaling methods discussed in this chapter. The UV approach is implemented in commercial software such as LISREL (Jöreskog and Sörbom 1993), Mplus (Muthén and Muthén 2000), and EQS (Bentler 1996).

7.11 Further reading

Bartholomew, D. J. and Knott, M. (1999). *Latent Variable Models and Factor Analysis*. 2nd edition. London: Arnold.

Bollen, K. A. (1989). *Structural equations with latent variables*. New York: Wiley and Sons.

Fischer, G. H. and Molenaar, I. W. (Eds.) (1995). *Rasch Models: Foundations, recent developments, and applications*. New York: Springer-Verlag.

Hambleton, R. K., Swaminathan, H. and Rogers, H. J. (1991). *Fundamentals of Item Response Theory*. Newbury Park, California: Sage Publications.

Heinen, T. (1996). *Latent Class and Discrete Latent Trait Models*. Thousand Oaks: Sage Publications.

van der Linden, W. J. and Hambleton, R. (1997). *Handbook of Modern Item Response Theory*. New York: Springer-Verlag.

CHAPTER 8

Factor Analysis for Ordered Categorical Variables

8.1 The practical background

The subject of this chapter lies, in a sense, between that of Chapter 6 on factor analysis and Chapter 7 on factor analysis for binary data. It differs from both of these earlier kinds of factor analysis principally in that the manifest variables are ordered categorical variables; that is, the response will be in one of a number of ordered categories. For example, if we ask someone whether they enjoyed a meal, "very much", "a little", or "not very much", we would be observing an ordered categorical variable. The answer can fall into only one category and those categories are ordered according to strength of approval. Any ordered categorical variable can be reduced, of course, to a binary variable by amalgamating categories. For example, if we were to amalgamate "a little" and "very much", we would have a binary variable. In fact, we did this in the analysis of the sexual attitudes data in Section 7.8, where three of the items had four categories. These were amalgamated into two pairs so that all variables could be treated as binary. In doing this, we are losing information and that provides the motivation for the present chapter. Categorical variables with more than two categories are often referred to as *polytomous* categorical variables.

In view of these considerations, it might have seemed logical to place this chapter immediately after the factor analysis chapter and then to treat the binary case in Chapter 7 as an important special case. However, work on ordered categorical variables is nearer to the research frontier and is consequently more incomplete and, in some senses, more difficult than the other methods. It is, therefore, better to approach it with the experience gained from the preceding two chapters on latent variable models. For the same reasons, the structure of this chapter will be a little different from its predecessors though we shall retain the emphasis on practical applications. The software available for these problems is more specialized and it will be necessary to give some account of what is currently available in Section 8.8.

In spite of all this, we think it is important for social science researchers to be aware of what is available because ordered categorical data are so common in social research. In many social applications, the data collected are coded into a number of ordered categories. Examples of ordered variables are often attitudinal statements with response alternatives such as "strongly disagree", "disagree", "agree" and "strongly agree" or "very satisfied", "satisfied", "dissatisfied" and "very dissatisfied". These scales are sometimes known as Likert-type scales. In such scales, it is also common to have an alternative which does not fit into the ordering. This is the case with responses such as "neither agree nor disagree", or "don't know". The treatment of scales including such items is discussed briefly in Section 8.6.

It will be useful to have an example in mind as we introduce the two main models on which the methods rest.

Example: Attitude to science and technology

The data used in this example come from the Consumer Protection and Perceptions of Science and Technology section of the 1992 Eurobarometer Survey (Karlheinz and Melich 1992) based on a sample from Great Britain. The questions chosen are given below.

1. Science and technology are making our lives healthier, easier and more comfortable. [Comfort]

2. Scientific and technological research cannot play an important role in protecting the environment and repairing it. [Environment]

3. The application of science and new technology will make work more interesting. [Work]

4. Thanks to science and technology, there will be more opportunities for the future generations. [Future]

5. New technology does not depend on basic scientific research. [Technology]

6. Scientific and technological research do not play an important role in industrial development. [Industry]

7. The benefits of science are greater than any harmful effects it may have. [Benefit]

All of the above items were measured on a four-point scale, with response categories "strongly disagree", "disagree to some extent", "agree to some extent" and "strongly agree".

Missing values have been excluded from the analysis by listwise deletion giving a sample of 392 respondents. Listwise deletion implies that response patterns that have missing values for any of the items are omitted from the analysis. Omitting respondents with missing values can bias the results.

To start our analysis, we chose the items that were positively worded namely Comfort, Work, Future, and Benefit. Those four items can be considered as indicators for measuring attitude towards science and technology.

8.2 Two approaches to modelling ordered categorical data

We have remarked that ordered categorical problems can be reduced to factor analysis of binary variables and the analysis of the sexual attitudes in Section 7.8 showed this being done. Dichotomization of ordinal variables is also applied to the science and technology data in Chapter 9. In the past, however, the approach has often been from the other end. Thus, in applications where the number of ordered categories is large (more than six or seven, say), ordinal categories have often been treated as if they were interval level variables. Having made that assumption, one can go on to compute correlations between these pseudo-continuous variables and carry out a standard factor analysis. Provided that the number of categories is large for all variables, this may not seriously affect the results of the analysis. Even when the number of categories is as low as three or four, it may be acceptable to use this method. We ourselves did this for the anxiety data in Chapter 6 though we emphasised that our concern there was to illustrate other aspects of factor analysis. In general, the uncritical factor analysis of categorical data in this way is likely to give biased estimates of the factor loadings and is not recommended. We shall provide empirical evidence of this point in Section 8.5.

One might still wonder whether factor analysis could be used on other types of correlation coefficient specifically designed for ordered categorical data. For example, Kruskal's *gamma*, Somer's *d*, or grouped forms of rank correlation coefficients such as Kendall's *tau* all measure the strength of the relationship between ordered categorical variables. Factor analyses are sometimes carried out on such coefficients and, if they are viewed from a purely descriptive point of view, they may yield useful insights. However, whatever the merits of such *ad hoc* methods, they have been superseded by better, model-based, methods.

As already explained in Chapter 7, there are two main approaches for analyzing binary data with latent variable models. Each of these can be generalized to the case of variables with more than two ordered categories. They are: the *item response function* (IRF) approach and the *underlying variable* (UV) approach. In this chapter, we will describe these two main methodologies used for multivariate analysis of ordinal items.

The IRF models use a straightforward extension of the logit or normit (probit) models for binary responses discussed in Chapter 7. It will be used in all the examples in this chapter. The logit model for ordinal responses is implemented in the software GENLAT provided on the Web site and MULTILOG (The program MULTILOG can fit a one-factor model and GENLAT up to two factors.)

The UV approach is based on the fit of the standard linear factor model using the matrix of polychoric correlation coefficients (which we explain below). This approach is supported by commercial software LISREL, EQS, and Mplus. The UV and IRF models will be compared through an example. More on the comparison between the two approaches can be found in two research papers by Moustaki (2000) and Jöreskog and Moustaki (2001).

8.3 Item response function approach

In the factor analysis for binary items, we were interested in modelling the probability of a randomly selected individual giving a positive response to an item as a function of the latent variables. This was done in terms of a set of probabilities $\{\pi_i(\mathbf{y})\}$. The probabilities of giving the negative response did not appear explicitly because they were simply the complements of the $\pi_i(\mathbf{y})$s. In the ordinal case, where there are more than two categories, we need to specify probabilities for each category. The observed ordinal variables are denoted by $x_1, \ldots, x_p$. Let us suppose that there are m_i categories for variable i labelled $(1, \ldots, m_i)$. For binary items, $m_i = 2$ for each i and the category labels were 0 and 1 but could equally well have been 1 and 2. We now need to define a response probability for each category. Let $\pi_{i(s)}(\mathbf{y})$ be the probability that, given $\mathbf{y}$, a response falls in category s for variable i.

The position with two categories can now be compared with the general case as follows:

Categories	0	1
Response probability	$1 - \pi_i(\mathbf{y})$	$\pi_i(\mathbf{y})$

Categories	1	2	$\ldots$	s	$\ldots$	m_i
Response Probability	$\pi_{i(1)}(\mathbf{y})$	$\pi_{i(2)}(\mathbf{y})$	$\cdots$	$\pi_{i(s)}(\mathbf{y})$	$\cdots$	$\pi_{i(m_i)}(\mathbf{y})$

In both cases, the response probabilities sum to one. In the binary case, we derived the logit model which expressed the logit of the probability of a response in category one as a linear function of the ys (see equation 7.3). The question now is how to generalize the argument used there to more than two categories. Suppose we were to divide the categories into two groups with categories $(1, 2, \ldots, s)$ in one group and $(s + 1, s + 2, \ldots, m_i)$ in the other and were merely to report into which of the two groups the response fell. We would thereby have reduced the polytomous variable to a binary variable. It therefore seems reasonable to require that any model we choose for the polytomous case should be consistent with the one which we have already used for the binary case. We can do this by supposing that *wherever we make the split* the binary logit model will apply. To do this, we need the probabilities of a response falling into the first and second groups, respectively. These may be written:

$$\gamma_{i(s)}(\mathbf{y}) = Pr(x_i \leq s) = \pi_{i(1)}(\mathbf{y}) + \pi_{i(2)}(\mathbf{y}) + \cdots + \pi_{i(s)}(\mathbf{y}),$$

and

$$1 - \gamma_{i(s)}(\mathbf{y}) = Pr(x_i > s) = \pi_{i(s+1)}(\mathbf{y}) + \pi_{i(s+2)}(\mathbf{y}) + \cdots + \pi_{i(m_i)}(\mathbf{y}),$$

where x_i denotes the category into which the ith variable falls.

The probabilities $\gamma_{i(s)}(\mathbf{y})$ are referred to as *cumulative response probabilities*. Note that it is only meaningful to describe them in this way because the categories are ordered.

In essence, we now define the model by supposing that the binary logit model holds for all possible divisions of the m_i categories into two groups. We can do this in two equivalent ways according to which group we regard as the "positive" response. That is, we can write the model in terms of $\text{logit}\gamma_{i(s)}(\mathbf{y})$ or of $\text{logit}(1 - \gamma_{i(s)}(\mathbf{y}))$. It would be natural to take the $(1 - \gamma)$ version because it links directly with the model for the binary case. The model is thus written

$$\log \left[\frac{1 - \gamma_{i(s)}(\mathbf{y})}{\gamma_{i(s)}(\mathbf{y})} \right] = \alpha_{i(s)} + \sum_{j=1}^{q} \alpha_{ij} y_j, \qquad (8.1)$$

where $(s = 1, \ldots, m_i - 1; \; i = 1, \ldots, p)$. This ensures that factor loadings have the same interpretation as in the binary case. In other words, the higher the value of an individual on the latent variable, the higher the probability of that individual belonging in the higher categories of an item. Instead of using the logit link function, we could use the normit (probit) function. The model that uses the logit as a link is also called the *proportional odds model*. The name proportional odds model comes from the fact that, in the one-factor case, the difference between two cumulative logits, that is, the left side of (8.1), for two persons with factor scores y_1 and y_2 is proportional to $y_1 - y_2$. Note that there is one intercept parameter $\alpha_{i(s)}$ for each category. This reflects the fact that as the threshold $(\alpha_{i(s)})$ rises for the response, the "difficulty" of the item will rise also. The ordering of the categories implies that the intercept parameters are also ordered, that is, either

$$\alpha_{i(1)} < \alpha_{i(2)} < \cdots < \alpha_{i(m_i)}$$

or, the reverse order applies. However, the factor loadings α_{ij} remain the same across categories of the same variable; in other words, the discriminating power of the item does not depend on where the split into two groups is made. The πs are obtained from the γs by

$$\pi_{i(s)}(\mathbf{y}) = \gamma_{i(s)}(\mathbf{y}) - \gamma_{i,(s-1)}(\mathbf{y}) \qquad (s = 2, \ldots, m_i), \qquad (8.2)$$

where $\gamma_{i(1)}(\mathbf{y}) = \pi_{i(1)}(\mathbf{y})$ and $\gamma_{i(m_i)}(\mathbf{y}) = 1$. We refer to $\gamma_{i(s)}(\mathbf{y})$ as the *cumulative response function* and to $\pi_{i(s)}(\mathbf{y})$ as the *category response function*.

We should mention that the regression model used for an observable ordinal response on a set of observable (rather than latent) explanatory variables is known as the *cumulative logit* model for ordinal variables.

Figures 8.1 and 8.2 give the cumulative and the category response functions respectively for parameter values $\alpha_{i(1)} = -0.5, \alpha_{i(2)} = 0.5, \alpha_{i(3)} = 1.5, \alpha_{i(4)} = 3.5$ and $\alpha_{i1} = 1.0$.

To summarize, the assumptions made under the IRF approach, which are common to other factor models, are

i) The latent variables are independent and normally distributed with mean zero and variance one.

ii) The responses to the ordinal items are independent conditional on the latent variables (conditional independence).

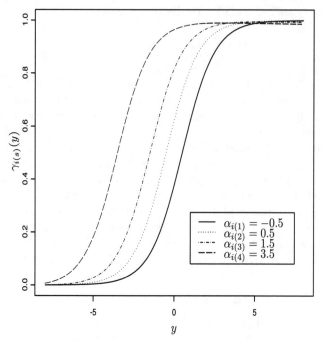

Figure 8.1 *Cumulative response probabilities*

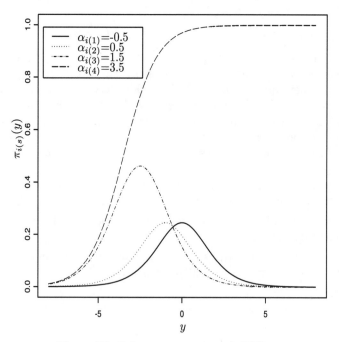

Figure 8.2 *Category response probabilities*

Fitting the model and goodness-of-fit

The model can be fitted in the same manner as the binary latent trait model using the method of maximum likelihood. A program for this called GENLAT is available on the Web site. It gives estimates of the parameters together with their asymptotic standard errors. As in the binary case, the input consists of the list of response patterns.

Goodness-of-fit can, likewise, be judged using the same criteria based on the likelihood ratio and the Pearson chi-squared statistics calculated from the whole response patterns. However, the problems of sparseness are liable to be much more serious in the polytomous case. We can easily see how this comes about by enumerating the number of response patterns. If there are m_i categories for variable i, the total number of response patterns is $(m_1 \times m_2 \times \cdots \times m_p)$. In the case of the science and technology data referred to at the beginning of the chapter, there were four categories for each variable giving $4^4 = 256$ response patterns altogether. This compares with 16 for binary data with four variables. The average expected frequency is therefore likely to be small even when the sample size is large. One way to combat this is to combine response patterns but there is another alternative which was not available for binary data. This is to amalgamate categories for some of the variables. It often happens that some of the response categories are rarely used and little is lost if we combine such a category with an adjacent category. This does not destroy the ordering but considerably reduces the number of response categories. We shall do precisely this when we come to the example on attitudes to the environment in the next section.

The number of degrees of freedom for G^2 or X^2 will be equal to the number of response patterns, after any grouping, less the number of independent parameters less one. Thus if there is no grouping, we have:

$$\text{degrees of freedom} = (m_1 \times m_2 \times m_3 \times \cdots \times m_p) - \sum_{i=1}^{p} m_i - p - pq - 1.$$

If there has been grouping of response patterns, the first term on the right will be reduced accordingly.

The goodness-of-fit can also be assessed by looking at the two-way (or higher) margins. The pairwise distribution of any two variables can be displayed as a two-way contingency table, and chi-squared residuals can be constructed in the usual way by comparing the observed and expected frequencies. We shall illustrate this in the following examples.

If we fit the one-factor proportional odds model to the science and technology data (the full data set is given on the Web site), we obtain the estimates given in Table 8.1.

The results suggest a strong common factor which loads heavily on all variables, but with the greatest weight on the "implications for the future". Even making allowance for the larger standard error of this loading, it seems clear that this difference is genuine. Before putting too much weight on this conclusion, we need to look at how well the model fits.

Table 8.1 *Estimated factor loadings with standard errors in brackets and standard-ized loadings for the one-factor model, science and technology data*

	$\hat{\alpha}_{i1}$	$\text{st}\hat{\alpha}_{i1}$
Comfort	1.04 (0.16)	0.72
Work	1.23 (0.19)	0.77
Future	2.28 (0.32)	0.92
Benefit	1.10 (0.16)	0.74

Given the sparsity of the data (there is a total frequency of 392 spread over 256 response categories), it is not feasible to carry out global tests. Instead we look at the fits to the margins. A fuller discussion of the use of margins to judge the fit of the proportional odds model is given in Jöreskog and Moustaki (2001). For each pair of items, Table 8.2 gives the sum of the chi-squared residuals over each pair of categories. Sixteen chi-squared residuals are calculated for each pair of items, since each variable has four response categories. Table 8.3 shows how the entry 25.54 of Table 8.2 is computed. The sum of the entries of Table 8.3 is 25.54. In a similar manner, we compute the sums of chi-squared residuals for the other five two-way tables.

Table 8.2 *Sums of chi-squared residuals for pairs of items from the two-way margins for the one-factor model, science and technology data*

	Work	Future	Benefit
Comfort	25.54	11.98	27.23
Work		9.21	23.27
Future			17.41

Table 8.3 *Chi-squared residuals for the two-way marginals of items Comfort and Work, science and technology data*

Categories	1	2	3	4
1	0.87	1.79	0.44	11.50
2	2.43	0.01	0.14	2.09
3	0.02	0.04	0.45	2.39
4	0.66	0.26	1.51	0.94

As discussed in Section 7.4, if the model were correct, the chi-squared residual in a single cell would have approximately a χ^2 distribution with one degree of freedom. A value greater than about 4 would indicate a poor fit. The sum of these residuals over all the cells in a two-way marginal table is analogous to Pearson's chi-squared statistic for goodness-of-fit, but because the model has been fitted to the full multi-way table (rather than to the two-way marginal table), the standard chi-squared test does not apply. We may still use this sum, say S, as a diagnostic guide. A large value of S would suggest that the associations in that two-way table are not well explained by the model. But how large is large?

In the absence of a general theoretical answer, we suggest two rules of thumb for an $(m_i \times m_j)$ marginal table. The first says that S is too large if it is greater than $(4 \times m_i \times m_j)$. This is adapted from Jöreskog and Moustaki (2001) which argues that since a chi-squared residual greater than 4 indicates a poor fit for a single cell, a value of S greater than $(4 \times m_i \times m_j)$ would indicate a poor fit for an $(m_i \times m_j)$ marginal table. In this book, we use a smaller value of $(3 \times m_i \times m_j)$.

The second rule says that S is too large if it is greater than the upper 1% point of a χ^2 distribution with $[(m_i \times m_j) - 1]$ degrees of freedom. The rationale behind this is that, had the model correctly specified the true cell probabilities for the two-way table, then S would have had this distribution, approximately. Because the cell probabilities are estimated (by fitting a latent variable model to the multi-way table) a value of S larger than this should be even more indicative of a bad fit.

Further work is needed to develop diagnostic procedures for deciding whether or not associations between pairs of items are well explained by a given latent variable model. However, in practice a model may still be useful even though it does not fully explain all interrelationships between items.

Returning to the science and technology data, all six entries, values of S, in Table 8.2 are less than both 48 (3×16) and 30.58 (the upper 1% point of χ^2 with 15 degrees of freedom). By either rule, the fit to each two-way marginal table appears satisfactory. However, in Table 8.3 one of the chi-squared residuals is quite large (11.50) even though their sum is only 25.54. But in all six two-way tables, there are only four chi-squared residuals greater than 3. These are listed in Table 8.4. Overall, the one-factor model appears to give an adequate, but not perfect, description of the data.

Table 8.4 *Chi-squared residuals greater than 3 for pair of items and categories for the one-factor model, science and technology data*

Items	Categories	$(O - E)^2/E$
(1,2)	(1,4)	11.50
(1,3)	(2,4)	6.80
(1,4)	(1,1)	16.46
(2,4)	(1,4)	13.23

Factor scores

It is possible to calculate factor scores for the proportional odds model, but much of the simplicity of the binary case is lost. In that case, we pointed out that we could either compute the conditional expectations of the latent variable, given the manifest variables, or we could use what we called components, which were linear combinations of the manifest variables formed using the factor loadings as weights. Either method ranked objects in the same order, and in practice, the results from the two methods were very similar. In the general case this simple correspondence breaks down. There are no components which convey all the information about the latent variables and we therefore have to rely on the conditional expectations which are more difficult to calculate.

8.4 Examples

Attitudes to the environment

As the first of our main examples, we take another data set on the environment, this time extracted from the Environment section of the 1990 British Social Attitudes Survey (Brook, Taylor, and Prior 1991). A sample of 291 individuals were asked whether they were "very concerned", "slightly concerned", "not very concerned", or "not at all concerned" with the following environmental issues.

1. Lead from petrol. [LeadPetrol]
2. River and sea pollution. [RiverSea]
3. Transport and storage of radioactive waste. [RadioWaste]
4. Air pollution. [AirPollution]
5. Transport and disposal of poisonous chemicals. [Chemicals]
6. Risks from nuclear power station. [Nuclear]

Since the proportion of individuals falling into the "not very concerned" category was less than 10%, categories "not very concerned" and "not at all concerned" were amalgamated. We treat the above six items as indicators of individuals' attitudes towards environmental issues. The one-factor model with the logit link function was fitted to the six items each with three ordered response categories. The parameters of interest are the factor loadings α_{i1} and their standardized form stα_{i1} and estimates of these are given in Table 8.5. The standardized parameters are all positive and close to one, indicating that the six items are all strong indicators of attitude towards the environment.

The goodness-of-fit of the model is again investigated through the chi-squared residuals computed for the two-way margins. The chi-squared residuals for the two-way margins are smaller than 3 for most of the cells. The sums of these residuals for the 15 two-way marginal tables are given in Table 8.6. These are smaller than both 27 (3×9) and 20.09 (the 1% point of χ_8^2), so both rules of thumb agree that the one-factor model adequately explains all the pairwise associations between the six attitude items. Therefore we can

Table 8.5 *Estimated factor loadings with standard errors in brackets and standard-ized loadings for the one-factor model, environment data*

Items	$\hat{\alpha}_{i1}$	st$\hat{\alpha}_{i1}$
LeadPetrol	1.38 (0.22)	0.81
RiverSea	2.35 (0.36)	0.92
RadioWaste	3.14 (0.42)	0.95
AirPollution	3.25 (0.49)	0.96
Chemicals	2.95 (0.43)	0.95
Nuclear	1.79 (0.28)	0.87

Table 8.6 *Sums of chi-squared residuals for pairs of items from the two-way margins for the one-factor model, environment data*

			Items		
Items	2	3	4	5	6
1	12.75	3.77	6.57	2.71	2.30
2		8.71	9.54	4.68	5.06
3			3.58	9.05	10.45
4				4.23	4.81
5					3.25

use this factor, with some confidence, as a summary measure of attitude to environmental issues.

Science and technology data

This is the same set of data that we used to introduce the chapter, but now we take all seven items from the Science and Technology section of the 1992 Eurobarometer Survey. This data set will also be analyzed in Chapter 9 in binary form using a three latent class model.

The first step is to fit a one-factor model to the seven items. The chi-squared residuals for the two-way margins are smaller than 3 for most of the cells but there are a few cells where the model shows a very bad fit. We report in Table 8.7 the sums of these chi-squared residuals for all pairs of items.

Values as big as 103.24, 98.36, and 90.76 indicate a very poor fit indeed. It is clear that we need to look at the two-factor model without inspecting the parameter estimates for the one-factor model.

The parameter estimates with the standard errors in brackets for the two-factor model are given in Table 8.8.

The two-factor model improves the fit of the two-way margins considerably, as we can see from Table 8.9. The sums of the chi-squared residuals for the

Table 8.7 *Sums of chi-squared residuals for pairs of items from the two-way margins for the one-factor model, science and technology data (seven items)*

Items	2	3	4	5	6	7
1	13.33	26.20	12.41	18.93	9.21	24.51
2		27.54	24.09	98.36	90.76	23.52
3			10.28	21.08	35.78	23.11
4				23.23	26.90	17.17
5					103.24	17.09
6						20.28

(Header spanning columns 2–7: Items)

Table 8.8 *Estimated factor loadings and standard errors in brackets for the two-factor model, science and technology data*

Items	Cat	$\hat{\alpha}_{i0(s)}$	$\hat{\alpha}_{i1(s)}$	$\hat{\alpha}_{i2(s)}$
Comfort	1	−5.00 (2.09)	0.27 (0.20)	1.16 (0.18)
	2	−2.74 (1.77)		
	3	1.53 (0.33)		
Environment	1	−3.45 (1.20)	1.61 (0.35)	0.09 (0.22)
	2	−1.26 (0.87)		
	3	0.99 (0.71)		
Work	1	−2.95 (0.86)	−0.39 (0.18)	1.20 (0.23)
	2	−0.90 (0.83)		
	3	2.30 (0.45)		
Future	1	−5.05 (1.92)	−0.30 (0.28)	2.16 (0.34)
	2	−2.13 (1.43)		
	3	1.90 (0.62)		
Technology	1	−4.17 (1.80)	1.71 (0.36)	0.08 (0.24)
	2	−1.49 (1.04)		
	3	1.07 (0.70)		
Industry	1	−4.71 (1.60)	1.55 (0.31)	0.55 (0.25)
	2	−2.53 (1.30)		
	3	0.45 (0.59)		
Benefit	1	−3.38 (1.46)	−0.08 (0.00)	1.12 (0.20)
	2	−1.00 (0.72)		
	3	1.71 (0.39)		

21 two-way marginal tables are all smaller than both 48 (3×16) and 30.58 (the 1% point of χ^2_{15}), thus the two-factor model appears to be adequate.

The two-factor solution obtained is not unique since orthogonal transformation of the factors leaves the value of the likelihood unchanged. However, in this particular example, the solution obtained gives two factors which can be interpreted without the need for rotation. The factor loadings of items

Environment, Technology, and Industry load heavily on the first factor and items Comfort, Work, Future, and Benefit on the second factor. It looks as if, on the one hand, people vary according to how important they judge science and technology to be, and on the other hand they vary in the extent to which they believe that technology can give answers to society's problems. An alternative explanation of why the items Environment, Technology, and Industry form a scale by themselves is that they are the ones that have been expressed negatively. This second explanation suggests that the result is an artefact of the questionnaire rather than of people's attitudes. The analysis carried out here agrees with the results found later when the latent class model with three classes is fitted to a dichotomized version of the seven ordinal items. We shall return to the interpretation of this set of data when we meet it again in Chapter 9.

Table 8.9 *Sums of chi-squared residuals for pairs of items from the two-way margins for the two-factor model, science and technology data*

Items	2	3	4	5	6	7
1	12.36	25.14	11.99	18.80	7.37	24.58
2		21.64	23.29	20.94	31.86	22.85
3			9.44	15.31	30.54	23.64
4				21.93	26.71	17.23
5					34.87	16.67
6						20.78

8.5 The underlying variable approach

The underlying variable (UV) approach follows the same lines as in the case of binary data. The essential assumption we made there was that we had a factor analysis problem in which we merely observed whether or not each variable exceeded a threshold. The problem then was how to carry out a linear factor analysis using the binary data.

When responses fall into m_i categories rather than two, we have rather more information about the incompletely observed variables. We now assume that the category in which a response occurs is determined by where the manifest variable falls in relation to a series of thresholds. Let us denote, as before, the observed ordinal variables by $x_1, x_2, \ldots, x_p$ and by $x_1^*, x_2^*, \ldots, x_p^*$ the incompletely observed variables or underlying variables. For variable x_i with m_i categories, there are $m_i - 1$ threshold parameters. Let the thresholds for variable i be denoted by,

$$\tau_{i(1)}, \tau_{i(2)}, \ldots, \tau_{i(m_i-1)}$$

This divides the scale of the underlying variable into m_i segments. The connection between the ordinal variable x_i and the underlying variable x_i^* is then that

$$x_i = s \quad \text{if} \quad \tau_{i(s-1)} < x_i^* \leq \tau_{i(s)} \qquad (s = 1, 2, \ldots, m_i).$$

The extreme lower and upper threshold parameters, $\tau_{i(0)}$ and $\tau_{i(m_i)}$, are $-\infty$ and $+\infty$, respectively.

Since ordinal information only is available about x_i^*, the mean and variance of x_i^* are not identified and are therefore set to zero and one, respectively.

The model to be fitted is the classical linear factor analysis model

$$x_i^* = \alpha_{i1}^* y_1 + \alpha_{i2}^* y_2 + \cdots + \alpha_{iq}^* y_q + e_i \qquad (i = 1, 2, \ldots, p), \qquad (8.3)$$

where e_i is a residual term and x_i^* is an unobserved continuous variable *underlying* the ordinal variable x_i. In classical factor analysis, x_i^* is directly observed but here it is only partially observed through x_i.

The reader who finds that the mathematical going is already becoming rather heavy needs only note at this point that if we could estimate the correlations between the underlying variables, x_i^*, we would have all that we needed to carry out a standard factor analysis. It is possible to estimate these correlations from the ordinal variables and the resulting estimates are called *polychoric* correlations. In the special case when there are only two categories, the corresponding coefficients are known, as noted in Chapter 7, as *tetrachoric* correlations. The "tetra" part refers to the four categories formed by cross-classifying two binary variables; the "poly" covers all cases where there are more than four such categories. For those who would like to go a little deeper into the method we give an outline in the following three paragraphs, but nothing essential will be missed if these are passed over.

In order to fit the model, it is clear that there are three different sets of parameters to be estimated, namely the thresholds, the polychoric correlations between the underlying variables, and, finally, the factor loadings of equation (8.3). Those three sets of parameters can in theory be estimated in one step. Under the assumptions of the linear factor model that the latent variables are independent and normally distributed with mean 0 and variance 1 and that the residual terms are independent and normally distributed with mean 0 and variance σ_i^2, it follows that the underlying variables $x_1^*, \ldots, x_p^*$ have a multivariate normal distribution with zero means, unit variances, and some (polychoric) correlation matrix $\mathbf{P} = \{\rho_{ij}\}$, where $\rho_{ij} = \sum_{l=1}^q \alpha_{il}^* \alpha_{jl}^*$. Although in principle we could estimate all the parameters simultaneously, the assumption of multivariate normality requires the evaluation of multiple integrals that make the estimation computationally infeasible when the number of ordinal variables is greater than about five.

One solution to that computational problem is to estimate the three sets of parameters in two or three stages. Software such as LISREL, Mplus, and EQS use a three-step procedure assuming underlying normally distributed variables x_i^*. In the first step, the thresholds are estimated from the univariate margins of the observed variables. In the second step, the polychoric correlations are estimated from the bivariate margins of the observed variables for given thresholds. In the third step, the factor model is estimated from the polychoric correlations by weighted least squares (WLS) using a weight matrix. The different weightings take into account the differing precisions of the correlation estimates. If one is only interested in obtaining consistent parameter estimates, then any positive definite weight matrix can be used. However, if one is interested in obtaining asymptotically correct chi-squared measures of goodness-of-fit and standard errors for the parameter estimates, then the correct matrix needs to be used. The asymptotic covariance matrix is often unstable in small samples and so it can only be used for large samples. When the sample size is small, "maximum likelihood estimation" is used instead. The assumptions made under the three-stage estimation procedure are:

i) The latent variables are independent and normally distributed with mean zero and variance one.

ii) The residuals are independent and normally distributed with mean zero and variance σ_i^2.

iii) Univariate and bivariate normality of the underlying variables x_i^*. The bivariate assumption can be tested using a large sample likelihood ratio or goodness-of-fit chi-squared test in each marginal two-way table. The software available provides those tests.

From the description of the three-stage estimation procedure, we can see that the model is not fitted to the whole response pattern (as is the case in the one-stage formulation) but rather to the univariate and bivariate distributions. For that reason, this approach is called a limited information method. The advantage of the three-stage method is that it is implemented in widely available software such as LISREL, Mplus, and EQS, and it allows researchers to fit a model with both a large number of ordinal items and a large number of latent variables.

In order to illustrate the results of calculating polychoric correlations and also to demonstrate that it is not sufficient to use product moment correlations calculated from the ordinal variables, Table 8.10 gives the Pearson correlation matrix and the polychoric correlations for four of the science and technology variables. The Pearson correlations are smaller than the polychoric correlations for all pairs of items. The polychoric correlations are better when the underlying variable model holds, and are generally larger than the Pearson product moment correlations.

Science and technology data

The one-factor model (equation (8.3) for $q = 1$) has been fitted to the polychoric correlation matrix of the four items from the science and technology data set using LISREL 8.5. The parameters of the model are estimated using two estimation methods, namely, maximum likelihood (ML) and weighted least squares (where the weight matrix is the inverse of the asymptotic covariance matrix of the polychoric correlations). Table 8.11 gives the estimated factor loadings $\hat{\alpha}_{i1}^*$ with standard errors in brackets and the chi-squared statistics

Table 8.10 *Correlation matrices for science and technology data*

	Pearson correlations				Polychoric correlations			
	Comfort	Work	Future	Benefit	Comfort	Work	Future	Benefit
Comfort	1.00				1.00			
Work	0.15	1.00			0.20	1.00		
Future	0.28	0.40	1.00		0.35	0.48	1.00	
Benefit	0.33	0.17	0.31	1.00	0.41	0.21	0.38	1.00

under the two estimation methods. The factor loadings for the two methods do not differ very much except for the item Comfort. The factor loadings are all positive, indicating a strong correlation between the items and the latent variable. The chi-squared statistic rejects the one-factor model under both ML and WLS. However, the WLS method gives a much smaller value than the ML method. As we have already argued, the chi-squared statistic should not be the only criterion for judging the goodness-of-fit of the model when, as in this case, the data are sparse. It would therefore be desirable to apply some of the other methods based on chi-squared residuals.

Table 8.11 *UV approach, estimated factor loadings with standard errors in brackets and chi-squared measures of fit, science and technology data (four items)*

	ML $\hat{\alpha}_{i1}^*$	WLS $\hat{\alpha}_{i1}^*$
Comfort	0.48 (0.06)	0.57 (0.07)
Work	0.55 (0.06)	0.56 (0.06)
Future	0.79 (0.06)	0.78 (0.06)
Benefit	0.51 (0.06)	0.58 (0.06)
chi-squared	28.77	10.21
p-value	0.00	0.0061

If we accept the fit as a first approximation to the truth, the conclusion to be drawn from fitting the UV model is essentially the same as that for the fit of the IRF model reported in Table 8.1. It supports the view that there is a single common factor underlying the data, of which the "Future" variable is the most important indicator. The obvious difference between the two analyses is that the loadings in the UV case are all smaller than for the logit IRF model. The factor loadings for the UV model are correlations between a normal latent variable and the normal underlying variables, whereas for the IRF logit model the standardized loadings are correlations between the normal latent variable and underlying variables that are not normally distributed. Had we used the normit IRF model instead of the logit, they would have been much closer.

The important point is that the *relative* values are much the same, and that is what matters for the purposes of interpretation. More information about the relationship between the numerical values of the loadings for the logit and normit models can be found in Bartholomew and Knott (1999).

Relationship between the UV and IRF approaches

Although the UV and the IRF models look quite distinct in the sense that the model fitting procedures and some of the model assumptions are different, an equivalence has been noticed between the two approaches by Bartholomew and Knott (1999). We made the same point in Chapter 7 when comparing the two approaches for binary data. The equivalence in the general case implies the following relationships between the parameters of the two models:

$$\alpha_{ij} = \frac{\alpha_{ij}^*}{\sigma_i} \, , \tag{8.4}$$

$$\alpha_{i(s)} = \frac{\tau_{i(s)}}{\sigma_i} \, , \tag{8.5}$$

where $\tau_{i(s)}$, α_{ij}^* and σ_i^2 are the thresholds, the factor loading of the jth latent variable and the variance of the error term in the linear factor model for the ith ordinal variable. The equivalence holds for the probit (normit) model and for the logit model in so far as the logit is a good approximation to the probit.

For the factor analysis model of (8.3), the correlation between a latent variable y_j and an underlying variable x_i^* is

$$\mathrm{Corr}(x_i^*, y_j) = \frac{\alpha_{ij}^*}{\sqrt{\sum_{j=1}^q \alpha_{ij}^{*2} + \sigma_i^2}} \, . \tag{8.6}$$

Substituting (8.4) into (8.6), we get the same correlation in terms of the IRF parameters α_{ij}. Those will be called standardized parameters and they will be denoted, as before, by stα_{ij}:

$$\mathrm{st}\alpha_{ij} = \frac{\alpha_{ij}}{\sqrt{1 + \sum_{j=1}^q \alpha_{ij}^2}} \, . \tag{8.7}$$

When results are presented from both the UV and the IRF approaches, it is best to produce the standardized version of the model parameters in order to allow comparisons.

In view of the equivalence between the two types of model, it is pertinent to ask why we have presented both and which is to be preferred in practice. We prefer the IRF version, and that is why we have used it for the analysis of the examples. It is a "full information" method in that the estimation method makes use of the full distribution over all score patterns. The UV method is a "partial information" method, using information only from the pairwise distributions of the ordinal variables. Our preference is therefore based partly on the grounds of efficiency. It is also based on the close link with the logit model for binary data which we have already given powerful reasons for adopting. There is also a link with the unordered model, which we shall look at briefly in Section 8.6. There is a comprehensive and appealing simplicity about including so much under a single umbrella.

Nevertheless, at the present stage, there are two arguments in favour of including the UV approach. It gives an alternative way of looking at the model which places it firmly in the factor analysis tradition. Given the relative newness of latent variable models for categorical data, it is useful to show that the novelty of such methods is not as great as appears at first sight. From this perspective, the method is merely a way of extending the scope of the factor model. Secondly, the software necessary to fit the model using polychoric correlations is widely available and familiar to social scientists. The balance of advantage may shift in the future and the provision of the software necessary to fit the IRF model provided with this book is a step in that direction.

8.6 Unordered and partially ordered observed variables

We have already remarked that categories such as "don't know" do not fit into a sequence of ordered categories obtained from attitude questions. Some method is therefore needed to deal with the case of partial ordering of categories. There are other cases where we suspect an ordering but would prefer the method to be flexible enough to allow us to check whether or not this is so. There are also many situations where some or all of the categorical variables are nominal. All of these cases can be handled by an alternative generalization of the binary latent trait model of Chapter 7. A full exposition of this model requires technical resources beyond the limits we have set ourselves for this book but a brief introductory account will help to complete the story of the factor analysis of categorical data.

The IRF approach which we used for binary data can easily be extended to cover the polytomous case. To do this, we treat each category of a nominal item as a binary item. An individual either belongs to that category or not. A binary response function model is then specified for each category of the nominal item having its own difficulty (intercept) and discrimination (slope) parameter.

We have already shown in Section 4.8 how to represent a polytomous variable by a vector of binary elements. In the present notation, the variable x_i is replaced by a vector-valued indicator function with its sth element defined as:

$$x_{i(s)} = \begin{cases} 1, & \text{if the response falls in category } s, \text{ for } (s = 1, \dots, m_i) \\ 0, & \text{otherwise} \end{cases}$$

where m_i denotes the number of categories of variable i and $\sum_{s=1}^{m_i} x_{i(s)} = 1$.

We now introduce a response function for each category of each variable exactly as in the ordinal case and using the same notation. Thus, the single response function of the binary case is now replaced by a set of response functions, one for each category, $\pi_{i(s)}(\mathbf{y})$ $(s = 1, \dots, m_i)$ where $\sum_{s=1}^{m_i} \pi_{i(s)}(\mathbf{y}) = 1$.

Constructing a model now requires us to express these response probabilities as functions of the latent variables. In Chapter 7, we explained why we could not choose a linear function and that led to the search for some function of the response probability which could be expressed as a linear function of the ys. Of the several possibilities, we favoured the logit function. The logit is the logarithm of the ratio of the response probability for one category to the corresponding probability for the other. In the polytomous case, it is not immediately clear how to extend this idea to the case of several categories. Whatever choice we make, the probabilities must add up to one across all categories. It turns out that this can be achieved if one category is designated as what we shall call the *reference* category and if the response probabilities for all other categories are expressed in terms of that probability. Therefore, we define $(m_i - 1)$ logits for pairs of categories where the reference category can be selected to be any of the m_i categories. If the first category of the nominal variable is selected to be the reference category, the generalized logit model for variable i may be written as:

$$\log \frac{\pi_{i(s)}(\mathbf{y})}{\pi_{i(1)}(\mathbf{y})} = \alpha_{i0(s)} + \alpha_{i1(s)}y_1 + \cdots + \alpha_{iq(s)}y_q , \qquad (8.8)$$

where $(s = 2, 3, \ldots, m_i)$. Equation (8.8) is the one used in multinomial logistic regression in the case where the variables on the right hand side of the equation are known. If we compare this with the binary model, the only difference on the right hand side is the subscript s appearing in all the parameters. The constants $\alpha_{i0(s)}$ are difficulty parameters not for the whole item i but for a specific category s of item i. The factor loadings or discrimination parameters, $\alpha_{ij(s)}$, measure the effect of the latent variable y_j on the log of the odds of being in category s rather than the reference category. We refer to this as the *nominal* model. For this purpose, it is convenient to re-analyze the environment data already given in Section 8.4. The results are given in Table 8.12.

Each variable has three categories and, since category 1 was chosen as the reference category, parameter estimates are given only for categories 2 and 3. The difficulty parameters $\alpha_{i0(s)}$ are all negative and fairly large, which means that the "unconcerned" category (category 3) has a very small chance of being chosen by the median individual in every case. The factor loadings given in the last column are the most interesting for the comparison with the proportional odds model. Although they have fairly large standard errors, it appears that in every case, the loading contrasting the first and the third category is greater than that contrasting the first and the second. The third category has a higher logit response probability and hence a higher probability of a positive response than does the second. Since the loadings for the second category are positive, we can infer that the second category has a higher probability of a positive response than the first category. In other words, the categories are ordered. Our analysis shows them to be ordered in precisely the same way as the ordinal analysis assumed them to be. Fitting the nominal model has confirmed the ordering of the categories. In this particular example, this is little more than

Table 8.12 *Estimated difficulty and discrimination parameters with standard errors in brackets for the one-factor model for nominal data, environment data*

Items	Category	$\hat{\alpha}_{i0(s)}$	$\hat{\alpha}_{i1}$
LeadPetrol	2	−0.75 (0.18)	1.34 (0.27)
	3	−3.06 (0.41)	2.06 (0.45)
RiverSea	2	−2.33 (0.37)	2.11 (0.44)
	3	−8.87 (2.60)	5.06 (1.74)
RadioWaste	2	−2.57 (0.50)	3.31 (0.73)
	3	−6.00 (1.16)	5.14 (1.12)
AirPollution	2	−1.39 (0.35)	3.10 (0.69)
	3	−8.53 (1.62)	6.52 (1.24)
Chemicals	2	−2.83 (0.62)	3.61 (0.95)
	3	−5.43 (1.08)	4.75 (1.20)
Nuclear	2	−0.33 (0.18)	1.54 (0.33)
	3	−1.95 (0.38)	2.82 (0.51)

proving the obvious but in less clear-cut examples, such confirmation would be useful.

The goodness-of-fit of the nominal model can also be investigated by computing the chi-squared residuals for the two-way margins. Table 8.13 gives the sums of the chi-squared residuals for the two-way marginal tables. Again the values are all smaller than both 27 (3×9) and 20.09 (the 1% point of χ_8^2). Judging by our two rules of thumb, the nominal model with one factor adequately explains the pairwise associations between the six items. Comparing the fit of the nominal model with the proportional odds model (Table 8.6), we see that for some pairs of items one model fits slightly better than the other. However, both models give a very good fit.

Table 8.13 *Sums of chi-squared residuals for pairs of items from the two-way margins for the one-factor model for nominal data, environment data*

Items	Items				
	2	3	4	5	6
1	13.46	4.95	7.92	2.51	1.69
2		5.02	8.38	3.76	4.05
3			4.47	8.17	12.96
4				2.06	6.03
5					6.08

In Chapter 7, we noted that for the binary response model the linear combinations of the manifest variables which we called components were "sufficient" in that they contained all the information in the data about the latent variable.

This property was lost when we moved to the ordinal model but it continues to hold for the nominal model. These "sufficient statistics" are weighted sums of the responses with weights equal to the discrimination coefficients: thus the score for the hth individual on the jth factor is

$$X_j = \sum_{i=1}^{p} \sum_{s=1}^{m_i} \alpha_{ij(s)} x_{ih(s)},$$

where $(j = 1, \ldots, q)$ and $x_{ih(s)}$ takes the value 1 when individual h responds to the sth category and 0 otherwise. These sufficient statistics, can be used as factor scores to scale individuals in the latent space.

The nominal model is the "natural" generalization of the binary model in the sense that most of the latter's attractive properties are preserved. It is broader than the ordinal (proportional odds) model in that it covers ordinal and nominal categories as well as mixtures of the two. It can also be used when missing values are present since "missing" can be included as a separate response category. The model will then predict the place of the missing value among the response categories. The model therefore lets the data "speak for itself" in a way that the ordinal model fails to do.

On the other side, there are two strong arguments for using the ordinal model whenever its assumptions are met. It makes use of relevant prior information about the ordering of the categories and when this is incorporated we might expect to get more precise estimates and more powerful tests of fit. More importantly, however, the flexibility of the nominal model is bought at a considerable price. The number of parameters to be estimated in the nominal model for variable i is:

$$(m_i - 1) + q \times (m_i - 1),$$

where $(m_i - 1)$ denotes the number of difficulty parameters and the number of loadings estimated under the nominal model which compares with

$$(m_i - 1) + q,$$

for the ordinal model. Both models involve a large number of parameters which can lead to numerical problems with the fitting but this is much worse for the nominal model. When, for example, $p = 7$, $q = 2$ and $m_i = 4$ for all i, there will be 63 parameters to estimate in the nominal model and 35 in the ordinal model. Often it will be found that the standard errors become very large and there are liable to be problems of identification. It is technical difficulties of this kind that make this more general approach hazardous.

8.7 Further examples and suggestions for further work

Government data

The data analyzed here relate to 786 respondents from the 1996 British Social Attitudes Survey (Jowell, Curtice, Park, Brook, and Thomson 1996). The data are given on the Web site.

The items selected for the analysis are:

On the whole do you think it should or should not be the government's responsibility to

1. provide a job for everyone who wants one, [JobEvery]
2. keep prices under control, [PriceCont]
3. provide a decent standard of living for the unemployed, [Living]
4. reduce income differences between the rich and the poor, [Income]
5. provide decent housing for those who can't afford it [Housing].

The four response alternatives are: "definitely should be", "probably should be", "probably should not be", and "definitely should not be".

You might expect that there would be a range of opinion on the extent to which government should provide for the basic needs of its citizens through social security so it is reasonable to try a model with one latent variable. The one-factor model for ordinal responses provides a good fit to the five items. Table 8.14 gives the sums of chi-squared residuals for the two-way margins and using our two rules of thumb these should be compared with either 48 (3×16) or with 30.58 (the 1% point of χ^2_{15}). One marginal table, cross-classifying item 1 by item 2 has a sum of chi-squared residuals, 40.24, lying between 30.58 and 48, all the others are smaller than both. There is some uncertainty as to the adequacy of the one-factor model. You could also repeat the analysis by fitting a latent class model as discussed in Chapter 9 to see whether the sample can be divided into two types of respondents. To perform the latent class analysis, you would first have to recode the variables as binary variables.

Table 8.14 *Sums of the chi-squared residuals for pairs of items from the two-way margins for the one-factor model for ordinal responses, government data*

	Item 2	Item 3	Item 4	Item 5
Item 1	40.24	22.87	7.87	14.42
Item 2		26.67	16.58	24.48
Item 3			9.66	17.96
Item 4				15.75

The estimated factor loadings for the one-factor model with their standard errors are given in Table 8.15. The standardized loadings are all close to one indicating a strong association between the latent variables and the five items.

Table 8.15 *Estimated factor loadings with standard errors in brackets and standardized loadings for the one-factor model, government data*

Items	$\hat{\alpha}_{i1}$	st$\hat{\alpha}_{i1}$
JobEvery	1.92 (0.14)	0.89
PriceCont	1.30 (0.12)	0.79
Living	2.65 (0.20)	0.94
Income	2.31 (0.19)	0.92
Housing	2.58 (0.23)	0.93

Voter's Study in Flanders, Belgium, 1991-1992

The set of questions analyzed here is from an opinion poll survey on political attitudes after general legislative elections (Carton, Swyngedouw, Billiet, and Beerten 1993). The survey was carried out between December 1991 and April 1992 in Flanders. The items chosen for analysis are expected to form a scale for measuring "Ethnocentrism". The following questions concern immigrants, by which we understand primarily Turks and Moroccans.

> Please tell me whether or not you agree with the following statements. Response alternatives: "completely agree", "agree", "neither agree nor disagree", "disagree", "completely disagree" and "no opinion".

1. Belgium shouldn't have brought in guest workers.
2. Generally speaking, immigrants can't be trusted.
3. Guest workers are a threat to the employment of Belgians.
4. Guest workers come here to exploit our Social Security.
5. In some neighbourhoods, government is doing more for immigrants than for the Belgians who live there.

The sample size was 2,691. "No opinion" is treated here as missing. The percentage of missing values for items 1, 2, 3, 4, and 5 is 2.1, 3.4, 2.0, 1.7, and 13.9 respectively. After missing values are eliminated using listwise deletion, we are left with 2227 respondents. A more correct analysis including all respondents and using a model that takes account of missing values was carried out by O'Muircheartaigh and Moustaki (1999). The results showed that missing values were not related to the "ethnocentrism" attitude, and so we believe that their elimination is not going to bias the results substantially. Table 8.16 gives the frequency distribution of the five items.

The response alternatives include a middle category "neither agree nor disagree", which might disrupt the ordering of the response categories. However, you could still start with five categories for each item by fitting the one-factor model for ordinal responses. The sums of chi-squared residuals are given in Table 8.17 and are all greater than both 75 (3×25) and 42.98 (the 1% point of χ^2_{24}) indicating that the one-factor model is a poor fit. The fit is not substantially improved when the two-factor model is fitted. Most of the large

Table 8.16 *Frequency distribution of observed items, Flander's data*

	Items				
Categories	1	2	3	4	5
1	0.15	0.08	0.10	0.20	0.16
2	0.24	0.18	0.28	0.37	0.32
3	0.22	0.29	0.25	0.21	0.24
4	0.34	0.37	0.33	0.18	0.25
5	0.05	0.07	0.05	0.04	0.04

chi-squared residuals are found between pairs of items for categories (1,2) and (4,5).

Table 8.17 *Sum of the chi-squared residuals for pair of items from the two-way margins for the one-factor model for ordinal responses, Flander's data*

	Item 2	Item 3	Item 4	Item 5
Item 1	171.33	169.04	161.66	149.48
Item 2		228.90	162.91	169.51
Item 3			142.93	195.18
Item 4				128.89

There are a number of possible reasons why the two-factor model for ordinal responses is not a good fit and these are given below.

i) The assumption of conditional independence is not satisfied and so more than two factors are needed to explain the associations among the five items.

ii) The logit model is not appropriate, a different link function is needed.

iii) The categories are not ordered. In that case, the model for partially ordered responses or nominal variables is likely to be more appropriate.

As far as point i) is concerned, a model with more than two factors cannot be fitted when there are only five items.

Instead of the two-factor logit model, the two-factor normit model could have been used but since the response functions are so similar in shape, no great differences are expected. You could try fitting one and two-factor models using the UV approach for ordinal response categories for comparison.

The point that remains to be investigated is the last one regarding whether the five response categories can be treated as ordered for each of the five attitude items. Close examination of the chi-squared residuals for each pair of categories (before they are added to produce the sum of residuals) suggests that problems occur for other categories as well as for "neither agree nor disagree". These chi-squared residuals are not given here but are available on the Web site, or you can create them yourself when you repeat the analysis.

The analysis is continued by fitting the model for nominal responses given in equation (8.8). The one-factor model was a bad fit judging by the large chi-squared residuals observed in the two-way margins. When the one-factor model for nominal items is fitted we would expect that, if the response categories are ordered, the factor loadings $\alpha_{i1(s)}$ should also be ordered across categories. This is because the factor loadings show the increase in the odds of falling into a category as the position of the individual on the latent variable increases. Remember that the first category of each item is treated as a reference category. The factor loadings are given in Table 8.18 and you can see that they are ordered across categories for all items, indicating that the middle category is in the middle of the scale for each item. However, when the five items were analyzed using the proportional odds model for ordinal responses, neither the one-factor nor the two-factor model was a good fit. You can also see that the factor loadings for the last two categories for each item are big, indicating that the response function fitted to those categories is very steep.

The two-factor model for nominal responses improved the fit in the margins. Table 8.19 gives the sums of chi-squared residuals for pairs of items. All sums are less than 75 (3×25), but two of them are greater than 42.98 (the 1% point of χ^2_{24}). The two rules of thumb disagree, but even if it is not a perfect fit, the model explains much of the two-way associations between the variables.

The estimated parameters of the two-factor model for nominal responses are given in Table 8.20. The nominal model produces an intercept and two factor loadings for each category of each item (excluding the reference category) whereas the ordinal model constrains all the factor loadings to be the same across categories. We therefore expect that the nominal model will fit better than the ordinal model since many more parameters are estimated under the nominal model. The nominal model makes the interpretation of the factor loadings more complicated since each item has, in our example, four factor loadings on each factor. On the other hand, the nominal model provides the researcher with a lot more information regarding the contribution of each category to the factors. The $\hat{\alpha}_{i1}$ coefficients from Table 8.20 are all positive, suggesting the existence of a general factor which contrasts "completely agree" for each item with the other four categories. The $\hat{\alpha}_{i2}$ loadings for all items are large for categories 4 and 5, indicating a steepness in the response function for these categories. Category 2 has coefficients close to zero or one for all items, indicating that the difference between the first and the second category is not substantial in discriminating between individuals. Based on that information we could group categories one and two together and categories four and five together without losing any important information.

From the analysis so far, we have concluded that the five items do not seem to be indicators of a single latent dimension called "ethnocentrism". On the other hand, when the two-factor model for ordinal responses was fitted, the fit was not improved. Also, not much information will be lost by grouping the

Table 8.18 *Estimated factor loadings with standard errors in brackets for the one-factor model for nominal responses, Flander's data*

Item	Category	$\hat{\alpha}_{i1(s)}$
1	2	1.49 (0.13)
1	3	2.24 (0.14)
1	4	2.95 (0.14)
1	5	4.35 (0.19)
2	2	2.54 (0.22)
2	3	3.63 (0.22)
2	4	5.02 (0.24)
2	5	6.71 (0.27)
3	2	3.87 (0.43)
3	3	5.12 (0.45)
3	4	6.32 (0.46)
3	5	8.18 (0.48)
4	2	2.69 (0.27)
4	3	4.73 (0.35)
4	4	6.73 (0.44)
4	5	10.00 (0.60)
5	2	1.90 (0.15)
5	3	2.35 (0.15)
5	4	3.55 (0.17)
5	5	5.60 (0.24)

Table 8.19 *Sums of the chi-squared residuals for pairs of items from the two-way margins for the two-factor model for nominal responses, Flander's data*

	Item 2	Item 3	Item 4	Item 5
Item 1	44.44	48.28	24.05	19.93
Item 2		34.88	19.42	21.64
Item 3			37.70	37.01
Item 4				18.79

first two and last two categories, and there is no evidence that the middle alternative is not in the middle of the response categories.

If you carry on the analysis by fitting a one-factor model for ordinal responses to the five items with three categories each, you will obtain the sums of the chi-squared residuals given in Table 8.21. Two out of these ten sums are greater than 27 (3×9) and four are greater than 20.09 (the 1% point of

Table 8.20 *Estimated difficulty and discrimination parameters with standard errors in brackets for the two-factor model for nominal responses, Flander's data*

Items	Category	$\hat{\alpha}_{i0(s)}$	$\hat{\alpha}_{i1}$	$\hat{\alpha}_{i2}$
1	2	1.69 (0.18)	2.23 (0.20)	0.05 (0.16)
1	3	1.73 (0.18)	2.68 (0.21)	0.73 (0.18)
1	4	2.03 (0.18)	2.59 (0.21)	1.60 (0.17)
1	5	−1.78 (0.37)	1.14 (0.29)	3.08 (0.27)
2	2	3.54 (0.46)	3.23 (0.33)	−0.05 (0.32)
2	3	4.68 (0.45)	3.83 (0.33)	1.13 (0.32)
2	4	4.67 (0.45)	4.00 (0.34)	2.90 (0.33)
2	5	0.70 (0.56)	2.83 (0.42)	4.95 (0.40)
3	2	4.72 (0.55)	3.67 (0.41)	1.11 (0.33)
3	3	4.93 (0.55)	4.37 (0.42)	2.22 (0.36)
3	4	4.83 (0.55)	4.52 (0.43)	3.47 (0.38)
3	5	−0.07 (0.74)	2.49 (0.56)	5.55 (0.49)
4	2	3.03 (0.33)	3.26 (0.34)	1.49 (0.30)
4	3	2.28 (0.34)	4.47 (0.41)	3.29 (0.40)
4	4	0.95 (0.39)	4.69 (0.45)	5.06 (0.48)
4	5	−3.82 (0.88)	2.09 (0.64)	6.97 (0.63)
5	2	2.13 (0.18)	2.26 (0.19)	0.69 (0.18)
5	3	1.90 (0.18)	2.44 (0.19)	1.14 (0.19)
5	4	1.46 (0.20)	2.82 (0.22)	2.33 (0.22)
5	5	−3.58 (0.61)	0.61 (0.39)	4.22 (0.41)

χ_8^2). Investigation of the chi-squared residuals across categories for pairs of items shows that the poor fit occurs in the middle category. Again, when the two-factor model is fitted, the fit is not improved considerably, indicating that the poor fit is not due to the existence of a second factor. There is no reason to suppose that a second factor is needed to explain the associations among the five items.

Table 8.22 gives the parameter estimates for the one-factor model for three ordinal response categories. The loadings are close to one, indicating a strong association between the single factor and each item.

This example illustrates the sad truth that statistical analyses do not always produce clear-cut results.

Table 8.21 *Sums of the chi-squared residuals for pairs of items from the two-way margins for the one-factor model for ordinal responses (three categories) , Flander's data*

	Item 2	Item 3	Item 4	Item 5
Item 1	24.46	28.00	17.07	21.32
Item 2		21.03	17.63	36.18
Item 3			24.00	13.90
Item 4				10.40

Table 8.22 *Estimated factor loadings with standard errors in brackets and standardized loadings for the one-factor model for ordinal responses (three categories), Flander's data*

Items	$\hat{\alpha}_{i1}$	st$\hat{\alpha}_{i1}$
1	1.53 (0.08)	0.84
2	2.17 (0.12)	0.91
3	2.09 (0.12)	0.90
4	3.07 (0.18)	0.95
5	1.49 (0.08)	0.83

8.8 Software

The UV models can be fitted using commercial software such as LISREL (Jöreskog and Sörbom 1993), EQS (Bentler 1996), and MPlus (Muthén and Muthén 2000). The IRF logit model with one latent variable can be fitted with the program MULTILOG (Thissen 1991) whereas for the case of one or two latent variables, the program GENLAT (Moustaki 2001) available on the Web site can be used.

8.9 Further reading

Bartholomew, D. J. and Knott, M. (1999). *Latent Variable Models and Factor Analysis*. 2nd edition. London: Arnold.

Krzanowski, W. J. and Marriott, F. H. C. (1995). *Multivariate Analysis, Part 2: Classification, Covariance Structures, and Repeated Measurements*. London: Arnold.

CHAPTER 9

Latent Class Analysis for Binary Data

9.1 Introduction

The situations in which we might wish to do a latent class analysis are very similar to those mentioned in connection with cluster analysis. Here, however, we shall usually have a clearer idea of how many clusters there might be and what they might represent. The following examples illustrate some possibilities.

i) *Educational assessment.* There has been a good deal of research on how children acquire new concepts. One hypothesis is that a child has either mastered the concept or has not. This may be judged by asking the child to perform tasks which depend for their successful completion on having grasped the concept. This is known as *criterion referenced* testing because performance is judged by reference to an external criterion. If the hypothesis is correct we might expect the "masters" to get all the items correct and the "non-masters" to get them all wrong. In practice, non-masters will sometimes get items correct by chance and masters will sometimes get items wrong by making silly mistakes. A latent class model may help us to decide whether or not the hypothesis is supported by the data allowing for any errors which have been made. In this case, we would fit a model with two classes - representing the "masters" and the "non-masters", respectively. An example of this type is discussed in Section 9.6 based on a data set analyzed in Macready and Dayton (1977).

ii) *Medical diagnosis.* Many medical conditions cannot be diagnosed directly, or without invasive surgery, because the root of the condition is deep-seated. However, many symptoms can be easily observed, some of which may point towards one cause and some to another. It would be useful if we could use observations of an individual's symptoms to estimate the probability that the patient has any of the possible conditions. A latent class model may help us to do this.

iii) *Selection methods.* Aptitude for performing a complex task, like flying an aircraft, can only be inferred in advance by testing the candidate's performance on a variety of tests designed to give an indication of the required skills. One might anticipate that there existed three classes: those who were ready for immediate acceptance, those who should be rejected, and those who, in time and with additional preparation, might make the grade. Latent class analysis would enable us to investigate this hypothesis and provide a rule for assigning individuals to classes.

Latent Class Analysis (LCA) is one of the most used of all latent variable methods. Its principal objectives are

i) To reduce the complexity of a data set by explaining the associations between the observed variables in terms of membership of a small number of unobservable latent classes, and hence to gain understanding of the interrelationships between the observed variables

ii) To be able to allocate an object to one of these classes — or sometimes just to estimate the probabilities of belonging to each class — on the basis of the values of the observed variables for that object

Latent class models assume conditional independence or local independence in the sense that, conditional on an object belonging to a given class, the observable variables are independent. The difference between latent class models and the factor analysis models discussed in the previous three chapters is that FA assumes that the latent variable (variables) is (are) metrical, and possibly normally distributed, whereas in LCA the *single* latent variable is categorical. In a model with J latent classes, the latent variable, y, can be defined to take the value 1 for an object in class 1, 2 for an object in class 2, ..., and J for an object in class J. (When $J = 2$, the classes might be labeled 0 and 1 rather than 1 and 2. In any case the precise labelling is irrelevant.) Although it is possible to fit latent class models for metrical (see Section 9.5) or for categorical manifest (observable) variables with more than two categories, we shall concentrate on the special case of LCA for binary manifest variables.

The structure of the attitude to abortion data analyzed using Cluster Analysis in Chapter 2 and again using Factor Analysis for binary data in Chapter 7 might have been better explained by a latent class model. The frequency distribution for the observed response patterns is given in Table 7.1. Response patterns (1111) and (0000) have the largest frequencies which suggests they might be used as the nuclei of two classes, one consisting of those tending to favour abortion and the other of those tending to oppose abortion. Part of the interest in trying to fit a latent class model is to see whether the remaining response patterns could be allocated to one or other of these classes. In that connection, it is interesting that 44 respondents had the pattern (0111), indicating that they agreed to all the items except the first one. The first item was the one found to have the largest "difficulty" coefficient in the latent trait analysis described in Chapter 7.

9.2 The latent class model for binary data

Suppose we have an n by p data matrix of values of p binary variables, $x_1, \ldots, x_p$ for n objects or individuals. In the attitude to abortion example, there are four binary variables or items and 379 individuals, or objects, who have responded to the items. The latent class model for binary variables with J latent classes makes the following assumptions:

i) The n objects are a random sample from some population and every object in that population belongs to just one of the J latent classes

ii) The probability of giving a positive response to a particular item is the same for all objects in the same class but may be different for objects in different classes

iii) Given the latent class to which an object belongs, its responses to different items are conditionally independent

Notation

Let $\pi_{ij} = \Pr(x_i = 1 \mid j)$ be the probability that a randomly selected object from class j will answer positively to item i, for $(i = 1, \ldots, p; \ j = 1, \ldots, J)$. Thus, π_{ij} is the conditional probability of a positive response to item i, given (or conditional on) membership of class j.

Let η_j be the proportion of the population in latent class j or equivalently the probability that a randomly selected object from the population belongs to latent class j, for $(j = 1, \ldots, J)$. Sometimes one refers to η_j as the *prior* probability of belonging to class j.

Note that in FA models, we conditioned on the values of q factors or latent traits, $y_1, \ldots, y_q$, and modelled the conditional distribution of $(x_1, \ldots, x_p)$ given $(y_1, \ldots, y_q)$. Here, we have a *single, categorical* latent variable y taking values $(1, \ldots, J)$ and by convention, we write π_{ij} or $\Pr(x_i = 1|j)$ rather than $\Pr(x_i = 1|y = j)$.

Fitting the model

The model is fitted iteratively to obtain maximum likelihood estimates, $\hat{\pi}_{ij}$ of π_{ij} and $\hat{\eta}_j$ of η_j. Unfortunately, the procedure may result in a local rather than a global maximum and there is no simple way to check whether a given solution is a true maximum or only a local one. We therefore recommend that the process is run several times with different starting points, hoping that if they all give the same solution, then it is likely to be the global maximum. Experience suggests that local maxima are unlikely to be a serious problem when only two or three classes are fitted, but that the risk is greater for larger numbers of classes.

There is another problem that can arise in fitting latent class models known as *under-identification*. Roughly speaking, this means that there are too many parameters to be estimated from the data available. An important difference between LCA and FA models for binary variable models is that the latter impose explicit functional relationships (such as the logit-linear relationship) between the probability of a correct/positive response and the latent variable or factor. The latent class models do not impose any such restriction on the form of the probabilities. This means that there are many more parameters to be estimated in a typical latent class model, and this has practical implications for the complexity of the models that can be fitted.

Interpretation of the latent classes

In some cases, we may have an idea about what classes we expect to find. We shall then wish to ask whether those uncovered by the analysis correspond to what we expected. In other cases, our approach is exploratory, and then we do not know in advance how many classes are needed or what they might represent. In the latter case, we usually start by fitting a model with two classes and then proceed to add further classes as necessary. In both approaches we need to be able to use the response probabilities for each class to infer something about the nature of the classes to which they relate. This process will be illustrated using the attitude to abortion data and other examples.

The estimates of the model parameters with their standard errors in brackets for the attitude to abortion data are given in Table 9.1. Individuals in Class 2 have much higher estimated probabilities of agreeing with all the propositions about abortion than those in Class 1. The estimated standard errors are sufficiently small for the probabilities to be taken at face value. Item 1, which says that the law should allow abortion when the woman decides on her own, has rather less support among members of Class 2 than other items ($\hat{\pi}_{12} = 0.71$).

Class 1 comprises individuals who have close to zero estimated probability of agreeing with item 1 and rather larger, though still small, estimated probabilities of agreeing with items 2, 3 and 4 ($\hat{\pi}_{21} = 0.09$, $\hat{\pi}_{31} = 0.12$, $\hat{\pi}_{41} = 0.15$). The more practical matters of marriage and affordability seem to be less critical in defining the attitudes of this class than the first two items. The last row of the table shows that about 39% of the individuals in the sample belong to Class 1 and 61% to Class 2.

Table 9.1 *Estimated conditional probabilities, $\hat{\pi}_{ij}$, and prior probabilities, $\hat{\eta}_j$, with standard errors in brackets for the two-class model, attitude to abortion data*

Item (i)	$\hat{\pi}_{i1} = \hat{\Pr}(x_i = 1 \mid 1)$	$\hat{\pi}_{i2} = \hat{\Pr}(x_i = 1 \mid 2)$
WomanDecide	0.01 (0.01)	0.71 (0.03)
CoupleDecide	0.09 (0.03)	0.91 (0.02)
NotMarried	0.12 (0.04)	0.96 (0.02)
CannotAfford	0.15 (0.04)	0.91 (0.02)
$\hat{\eta}_j$	0.39 (0.03)	0.61 (0.03)

Goodness-of-fit

The three methods of assessing goodness-of-fit described in Chapter 7 on FA for binary data can also be applied to latent class models. The only new feature concerns the degrees of freedom for the two goodness-of-fit statistics, G^2 and X^2. The degrees of freedom equal the number of different response patterns (2^p) minus the number of independent parameters ($J - 1 + Jp$) plus one which is equal to $(2^p - J(p + 1))$. For a test to be possible, the degrees of freedom must be greater than zero. Problems will arise if p is small and J is large. For example, for the attitude to abortion data $p = 4$, so when $J = 2$ there are six degrees of freedom, but this will be reduced if response patterns have to be grouped to ensure that the expected frequencies are not less than 5. For the attitude to abortion data, the log-likelihood-ratio statistic is $G^2 = 37.02$ and the chi-squared test statistic is $X^2 = 44.81$ on six degrees of freedom, indicating that the two-class model is not a good fit to the data. The values of G^2 and X^2, after grouping the response patterns with expected frequency less than 5, become 34.84 and 31.09, respectively, on one degree of freedom, which also indicates a poor fit.

The chi-squared residuals for the attitude to abortion data given in Table 9.2 show that the two-class model predicts the two-way margins very well for most pairs of items. There is only one pair with a large residual for response (0,1) to items 2 and 1. Judging the model by how well it predicts the two- and three-way margins, we have no reason to reject the two-class model. The chi-squared residuals for the (1,1,1) three-way margins are all close to zero.

The percentage of G^2 explained is 94.2% indicating that the two-class model is a much better fit than the independence model. If we decide that the fit is not good enough we could go on to fit a three-class model, but it is here that we run into trouble with identifiability. When $J = 3$, the number of degrees of freedom becomes one and these may be lost if grouping takes place. With more than three classes, there will certainly not be enough degrees of freedom to make a test. We therefore have to do the best we can with a less well fitting model. In the case of the attitude to abortion data, it is possible to make a sensible interpretation of the data using the two-class model (see Table 9.3), even though the requirement of conditional independence (assumption iii) may not be fully satisfied.

Allocation to classes

Having decided to use the latent class model as a useful simplification of our data, we wish to allocate the objects to the identified classes using their responses. This is equivalent to the "factor scores" problem in factor analysis and latent trait analysis. The question is: what can we say about the class membership of the objects after they have responded to the items. We solve the problem by estimating the probability that an object with a particular response pattern falls into a particular class. This probability, sometimes called the *posterior* probability, is:

$$\Pr(\text{object is in class } j \mid x_1, \ldots, x_p) \qquad (j = 1, \ldots, J).$$

Table 9.3 gives the estimated allocation probabilities for the attitude to abortion data. Most of the probabilities are close either to zero or one, indicating that there is little doubt as to the class to which each individual should be allocated. In a few cases, the position is more ambiguous, especially for response pattern 0101 where the probability is 0.55. However, there are

Table 9.2 *Chi-squared residuals for the second order margins for a two-class model, attitude to abortion data*

	Items	O	E	$O - E$	$(O - E)^2/E$
Response (0,0)	2, 1	147	137.79	9.21	0.62
	3, 1	131	130.16	0.84	0.05
	3, 2	117	117.58	−0.58	0.00
	4, 1	129	129.12	−0.12	0.00
	4, 2	114	114.61	−0.61	0.00
	4, 3	116	109.97	6.03	0.33
Response (0,1)	1, 2	66	75.21	−9.21	1.13
	1, 3	82	82.84	−0.84	0.01
	1, 4	84	83.88	0.12	0.00
	2, 1	7	16.21	−9.21	5.24
	2, 3	37	36.42	0.58	0.01
	2, 4	40	39.39	0.61	0.01
	3, 1	7	7.84	−0.84	0.09
	3, 2	21	20.42	0.58	0.02
	3, 4	22	28.03	−6.03	1.30
	4, 1	16	15.88	0.12	0.00
	4, 2	31	30.39	0.61	0.01
	4, 3	29	35.03	−6.03	1.04
Response (1,1)	2, 1	159	149.79	9.21	0.57
	3, 1	159	158.16	0.84	0.01
	3, 2	204	204.58	−0.58	0.00
	4, 1	150	150.12	−0.12	0.00
	4, 2	194	194.61	−0.61	0.00
	4, 3	212	205.97	6.03	0.18

only six respondents in this category. Most individuals can be allocated to one or other latent classes with little uncertainty. The last column gives the allocation of response patterns to clusters 1 and 2 based on a cluster analysis performed in Chapter 2. The allocation based on LCA and CLA do not match exactly. Which do you think best?

Table 9.3 *Estimated posterior probabilities of class membership for the attitude to abortion data*

Response pattern	$\hat{\mathrm{Pr}}(\text{Class } 1 \mid x_1, \ldots, x_p)$	$\hat{\mathrm{Pr}}(\text{Class } 2 \mid x_1, \ldots, x_p)$	Class allocation	CLA cluster allocation
0000	1.00	0.00	1	1
0001	0.99	0.01	1	1
0010	0.96	0.04	1	2
0100	0.98	0.02	1	1
1000	0.95	0.05	1	1
0011	0.31	0.69	2	2
0101	0.45	0.55	2	1
0110	0.21	0.79	2	2
1100	0.16	0.84	2	1
0111	0.00	1.00	2	2
1011	0.00	1.00	2	2
1101	0.00	1.00	2	1
1110	0.00	1.00	2	2
1111	0.00	1.00	2	2

9.3 Example: attitude to science and technology data

The seven variables used in this example have already been given in Chapter 8 in their original form. Originally, all the items were measured on a four-point scale with response categories "strongly disagree", "disagree to some extent", "agree to some extent", and "strongly agree". For the purpose of the present analysis, response categories were dichotomised as follows: categories "strongly disagree" and "disagree to some extent" coded as 0 and categories "agree to some extent" and "strongly agree" as 1. The items Environment, Technology, and Industry are negatively expressed. These items have been recoded so that a high score (1) corresponds to a positive attitude towards science and technology.

The number of respondents who answered all seven items is 392. There are $2^7 = 128$ possible response patterns so each pattern occurs, on average, about three times, but there is a great deal of variation and many patterns do not occur at all. The full data set is given on the Web site.

First we fit the two-class model. Table 9.4 gives the estimated probabilities and their asymptotic standard errors for the two-class model.

Class 2, containing an estimated 79% of the population, consists of those individuals who are likely to have a positive attitude to all seven items. This may be described as a pro-science class. The way to describe Class 1 is less clear. On some items, Environment, Technology, and Industry, the probabilities are very similar to those for Class 2 so one cannot describe Class 1 as anti-science. They are, however, markedly less likely to respond positively about the prospects for Work, the Future, and about the benefits outweigh-

Table 9.4 *Estimated conditional probabilities, $\hat{\pi}_{ij}$, and prior probabilities, $\hat{\eta}_j$, with standard errors in brackets for the two-class model, science and technology data*

Item (i)	$\hat{\pi}_{i1} = \hat{\Pr}(x_i = 1 \mid 1)$	$\hat{\pi}_{i2} = \hat{\Pr}(x_i = 1 \mid 2)$
Comfort	0.75 (0.06)	0.95 (0.02)
Environment	0.76 (0.06)	0.68 (0.03)
Work	0.36 (0.09)	0.75 (0.04)
Future	0.20 (0.18)	0.94 (0.04)
Technology	0.74 (0.06)	0.72 (0.03)
Industry	0.84 (0.05)	0.86 (0.02)
Benefit	0.41 (0.09)	0.77 (0.03)
$\hat{\eta}_j$	0.21 (0.07)	0.79 (0.07)

ing the harmful effects of science. Thus, whereas all tend to agree about the "technical" advances of science, Class 1 members are more skeptical about the social benefits. Although this seems a reasonable way of interpreting the analysis, the fit of the two-class model is, in fact, not very good, at least as judged by the global tests. The X^2 and G^2 statistics after grouping the response patterns with expected frequencies less than 5 are 118.48, and 109.03, respectively, on four degrees of freedom. Both measures indicate a very poor fit. The percentage of G^2 explained by the two-class model is 27.7%. Table 9.5 gives the chi-squared residuals for pairs of items where the residual was greater than 3. There was also a large residual, 5.19, for the positive responses to items 2, 5, and 6 among the three-way margins. As already mentioned in Chapter 7, we use as a rule of thumb a value of the residuals greater than 4 as an indication of poor fit. However, since that value is only indicative in the examples, we also report residuals greater than 3.

The fit revealed by the margins suggests that the model captures an important part of the data structure. It also suggests that something more might be learnt by looking at a three-class model.

The parameter estimates for the three-class model are given in Table 9.6. Class 3 remains more or less the same as Class 2 from the two-class solution. This is true both of its size (79%) and the response probabilities. Asymptotic standard errors cannot be computed because some of the estimated conditional probabilities are close to the boundaries.

Individuals in Class 3 tend to have strong positive attitudes towards science and technology, and we may continue to refer to them as the pro-science group. What appears to have happened in the three-class solution is that the original Class 1 is now split into two separate classes. The new Class 1 is much the same as Class 1 in the two-latent class solution, except that it is now much smaller (9%). The main difference is that respondents have almost a zero probability of agreeing that the benefits of science outweigh the harmful effects. The new Class 2 is distinguished by a low or zero probability of agreeing to Environment and Technology items, but on all items they differ

Table 9.5 *Chi-squared residuals greater than 3 for the second order margins for the two-class model, science and technology data*

	Items	O	E	$O - E$	$(O - E)^2/E$
Response (0,0)	5, 1	17	10.05	6.95	4.81
	5, 2	52	33.18	18.82	10.67
	6, 1	10	5.56	4.44	3.54
	6, 2	31	17.23	13.77	11.00
	6, 5	26	15.83	10.17	6.53
Response (0,1)	2, 5	67	85.82	−18.82	4.13
	5, 2	57	75.82	−18.82	4.67
	6, 2	26	39.77	−13.77	4.77
	6, 5	31	41.17	−10.17	2.51
Response (1,0)	2, 5	57	75.82	−18.82	4.67
	2, 6	26	39.77	−13.77	4.77
	5, 2	67	85.82	−18.82	4.13
	5, 6	31	41.17	−10.17	2.51

from both the other classes. Similar patterns were found when items were analyzed as four-point scales with ordinal items as we saw in Chapter 8.

Table 9.6 *Estimated conditional probabilities, $\hat{\pi}_{ij}$, and prior probabilities, $\hat{\eta}_j$, for the three-class model, science and technology data*

Items	$\hat{\pi}_{i1}$	$\hat{\pi}_{i2}$	$\hat{\pi}_{i3}$
Comfort	0.63	0.80	0.95
Environment	0.74	0.29	0.76
Work	0.28	0.92	0.67
Future	0.16	0.95	0.82
Technology	0.74	0.00	0.83
Industry	0.74	0.55	0.92
Benefit	0.00	0.72	0.77
$\hat{\eta}_j$	0.09	0.12	0.79

The overall goodness-of-fit measures could not be used here because, after grouping, there were no degrees of freedom left for testing the fit.

The fit of the model judged by the chi-squared residuals on the two-way margins and the (1,1,1) three way margins is greatly improved. There are no discrepancies greater than 1 for any pair or triples of items.

The amount of G^2 explained increases from 27.7% to 48.6% as we go from two to three classes.

It therefore appears that the two-class solution did convey the essence of the situation but that the three-class solution has enabled us to describe the minority group in more detail.

The sexual attitudes data revisited

The questions on sexual attitudes from the 1990 British Social Attitudes Survey were analyzed in Chapter 7 using a two-factor latent trait model. We found that the two-factor solution was unsatisfactory both in terms of fit and by failing to produce two clear-cut factors. Here we re-analyze the data using a latent class model. The same set of questions has been thoroughly discussed in de Menezes and Bartholomew (1996).

Latent class models with two, three, and four classes were fitted to the ten items. The likelihood ratio and chi-squared test statistics after grouping for the three models are given in Table 9.7. None of the models gives a satisfactory fit judging by the overall measures of goodness-of-fit. Note that no degrees of freedom were left after grouping in the four-class solution.

Table 9.7 *Goodness-of-fit measures, sexual attitudes data*

	two-class	three-class	four-class
Likelihood ratio	705.01	327.45	302.67
chi-squared	1002.92	268.80	255.47
df	18	7	–

The parameter estimates for the two, three, and four class solution are given in Table 9.8.

Table 9.8 *Estimated conditional probabilities, $\hat{\pi}_{ij}$, and prior probabilities, $\hat{\eta}_j$, sexual attitudes data*

Item	Two-class		Three-class			Four-class			
	$\hat{\pi}_{i1}$	$\hat{\pi}_{i2}$	$\hat{\pi}_{i1}$	$\hat{\pi}_{i2}$	$\hat{\pi}_{i3}$	$\hat{\pi}_{i1}$	$\hat{\pi}_{i2}$	$\hat{\pi}_{i3}$	$\hat{\pi}_{i4}$
1	0.13	0.12	0.13	0.09	0.21	0.14	0.09	0.21	0.07
2	0.76	0.88	0.76	0.87	0.92	0.77	0.87	0.92	0.60
3	0.64	0.88	0.64	0.86	0.96	0.63	0.86	0.96	0.87
4	0.09	0.17	0.09	0.13	0.30	0.08	0.13	0.31	0.27
5	0.08	0.49	0.08	0.38	0.82	0.07	0.38	0.83	0.60
6	0.01	0.92	0.01	0.87	0.97	0.01	0.87	1.00	0.00
7	0.83	0.99	0.06	0.98	1.00	0.06	0.98	1.00	0.20
8	0.23	0.94	0.21	0.91	0.99	0.21	0.91	1.00	0.27
9	0.07	0.30	0.07	0.10	0.98	0.05	0.11	0.98	1.00
10	0.02	0.19	0.02	0.00	0.84	0.00	0.00	0.85	1.00
$\hat{\eta}_j$	0.49	0.51	0.47	0.41	0.11	0.46	0.42	0.11	0.01

The two-class solution divides the population into two groups of similar size (49 and 51%). Individuals in Class 1 have low probabilities of agreeing with any of the ten items except items 2 and 3. Members in Class 2 have high probabilities of agreeing with all ten items. The two-class solution provides a permissive/nonpermissive dichotomization of the population. However, the two-class solution was not a good fit. We proceed to the three-class solution which has the first class to be the same in terms of size and response probabilities as Class 1 of the two-class solution. Class 2 remains, to some extent, as the permissive one, with low probabilities on the adoption items 9 and 10. Class 3 comprises 11% of the population, and members in this class have the highest probability of agreeing for each item. In the four-class solution, the first two classes remain the same as the first two classes in the three-class solution. The 11% in the third class is now split into two classes. One is identical to the third class of the three-class solution and the other one consists of only 1% of the population. That class reveals a completely different group of individuals showing a positive attitude towards the adoption items but opposing gay people teaching in schools and universities and holding public positions.

In the sexual attitudes example, the latent class analysis is more revealing than the factor analysis.

9.4 How can we distinguish the latent class model from the latent trait model?

One reason for using the attitude to abortion data to illustrate both the latent trait model and the latent class model was to show, incidentally, that the two models are often equally successful in fitting a set of data. In this particular case, neither model provided a particularly good fit but both were good enough to provide reasonable interpretations. The results for the attitude to abortion data are given in Table 9.9.

In other examples where the fit is much better, one often finds that the expected frequencies for the two models are virtually identical. At first sight, it seems surprising that models which make such different assumptions about the distribution of the latent variable should give such similar fits. In the case of the latent trait model, the latent variable is assumed to have a standard normal distribution. We have already remarked that the choice of the form of this distribution is not critical. It now appears that even if we replace the continuous prior by a distribution of the latent variable which is concentrated on a small number of discrete points (which is what the latent class model amounts to), the fit is not much affected. Another way of putting this is to say that the distribution of the latent variable is poorly determined by the data. What, then, are we to do about interpretation when our various methods for judging fit have little chance of distinguishing between the models?

Two things may be said. First, we should be wary of pressing our conclusions to the point where the distinction between the models has practical implications. For example, our analysis of the attitude to abortion data makes it very clear that people differ in their attitude to abortion. It is less clear

Table 9.9 *Observed and expected frequencies under the latent trait and the latent class models, attitude to abortion data*

Observed	Expected latent trait	Expected latent class	Response pattern
103	100.0	98.0	0000
13	16.6	17.8	0001
1	1.7	1.1	1000
9	9.1	10.3	0100
10	12.3	14.2	0010
6	7.4	4.1	0101
3	1.0	0.6	1100
21	14.8	7.8	0011
7	12.3	6.7	0110
3	1.9	5.4	1101
6	6.2	13.0	1011
44	41.1	54.2	0111
12	7.2	12.8	1110
141	143.9	130.9	1111

whether that variation is best described by a distribution of attitudes along a continuum or by a polarisation into two groups. There would therefore be little empirical justification for claiming that about 61% of the population were pro-abortion, because that depends on the selection of the latent class model as the more appropriate. It might, however, be the appropriate model to use if one wanted to predict the outcome of a vote on the question since that would force people to adopt one or other stance.

Secondly, background knowledge, which cannot easily be quantified, may favour one model over the other. The sexual attitudes data illustrates this point. Having fitted a model with two continuous factors, we still did not have a very satisfactory fit although it was clear that the latent trait model did capture some important aspects of the variation between individuals. The latent class model seemed to be more successful in providing an interpretable solution by identifying minority groups whose existence seemed credible on general grounds.

The relationship between the latent trait and latent class models can be explored theoretically, and the interested reader may refer to Bartholomew and Knott (1999), p.135-137. There is one aspect of this comparison which we mention here because it links back to the "threshold" effect which we met in Chapter 7. We noted there that estimated values of the discrimination parameter α_{i1} were sometimes very large. This implied that the item response function was almost vertical at its central point, which means that the probability of a positive response switches from a low value to a high value at that point. This is why it is called a threshold effect. But this is exactly what happens with the two-class latent class model — as we move from one class to the other, the probability of a positive response also changes abruptly. The appearance of large discrimination parameters in a latent trait model is therefore an indication that a latent class model might be more appropriate.

9.5 Latent class analysis, cluster analysis, and latent profile analysis

Latent class analysis may be regarded as a form of cluster analysis. It differs from the methods given in Chapter 2 principally in that it starts from a probability model. It is worth spelling out some of the similarities and differences. CLA begins by constructing a similarity (or distance) matrix and then goes on to look for clusters of objects which are close together. LCA begins with the hypothesis that there are J clusters and constructs a method whereby objects may be allocated to clusters. Unlike CLA, this allocation is probabilistic; instead of knowing which cluster an object belongs to, we merely have probabilities of belonging to the various clusters. The rationale behind CLA is rooted in the similarities between rows of the data matrix; LCA is based on the probabilities of the elements in the rows. Thus, it sets out to construct clusters all of whose members have the *same* probabilities of yielding a positive response. (This means that the rows of the data matrix within a cluster have the same expectation.) In order to achieve this, it assumes that, within those clusters, the outcomes are independent. In the examples we have met, CLA involved fairly small sample sizes and (where they were used directly) large numbers of variables. LCA, typically, has used large sample sizes and small numbers of variables. However, this is not a fundamental distinction.

If we return to Table 6.1 in Chapter 6, there is one cell of the table, Latent Profile Analysis, which has not been mentioned up to this point. Latent profile analysis differs from latent class analysis only in that the manifest variables are metrical instead of binary. This is a topic which has not received as much attention as the others covered in this book, and there is good reason why its use is problematical. We have noted above how difficult it is to distinguish empirically between a latent trait and a latent class model. The situation is even worse when we come to factor analysis and latent profile analysis. In this case, it is virtually impossible to distinguish between the two models because both models have essentially the same correlation structure. By this we mean that, given any correlation matrix, we can find a factor model and a latent profile model which would be equally successful in producing that matrix. On the basis of the correlation matrix, there is therefore no means of distinguishing between them. A distinction can be made only on non-empirical grounds. This is not to say that latent profile models should not be used. They may provide a more natural interpretation, for example when used to detect possible clustering. Perhaps the best way to approach the issue is through factor analysis, where we can ask whether a given factor solution might be better approached from a latent profile perspective. This is a matter for future research, and we leave the matter there.

9.6 Further examples and suggestions for further work

Three more data sets will be offered for analysis in this section, of which two have been analyzed in Chapter 7 using the latent trait model for binary responses.

Macready and Dayton data

This data set arises from educational testing where one wishes to study the learning process in children. Macready and Dayton (1977) used a two-class model on a set of four tests to identify two distinct classes; that of "masters" and "non-masters". The frequency distribution of the response patterns together with the expected frequencies for each response pattern under the two-class model and the probability of belonging to the "master" class are given in Table 9.10. The estimated posterior probabilities show that most of the individuals can be allocated to the "master" class or the "non-master" class with high confidence. For example, individuals who have responded to at least two items correctly have probabilities greater than 0.9 of being allocated into the "master" class. On the other hand, individuals who got all items wrong (0000) are allocated to the "non-master" class with probability 0.98. There are, however, three response patterns (0100, 0001, 0010) that have probabilities close to a half. These include only 11 individuals out of 142. Therefore, the latent class model has been able to classify with high confidence most of the response patterns in the two classes.

Table 9.10 *Observed and predicted frequencies and estimated class probabilities for the two-class model, Macready and Dayton data*

Observed frequency	Expected frequency	$\hat{\text{Pr}}(\text{master} \mid \mathbf{x})$	Class	Response pattern
15	14.96	1.00	2	1111
23	19.72	1.00	2	1101
7	6.19	1.00	2	1110
4	4.90	1.00	2	0111
1	4.22	1.00	2	1011
7	8.92	0.91	2	1100
6	6.13	0.90	2	1001
5	6.61	0.98	2	0101
3	1.93	0.90	2	1010
2	2.08	0.97	2	0110
4	1.42	0.97	2	0011
13	12.91	0.18	1	1000
6	5.62	0.47	1	0100
4	4.04	0.45	1	0001
1	1.31	0.44	1	0010
41	41.04	0.02	1	0000

The $X^2 = 9.5$ and the $G^2 = 9.0$ on six degrees of freedom indicate a near perfect fit to the data. The percentage of G^2 explained is 91%. The parameter estimates and standard errors of the model are given in Table 9.11.

Table 9.11 *Estimated conditional probabilities, $\hat{\pi}_{ij}$, and prior probabilities, $\hat{\eta}_j$, with standard errors in brackets for the two-class model, Macready and Dayton data*

Item (i)	$\hat{\pi}_{i1}$	$\hat{\pi}_{i2}$
1	0.21 (0.06)	0.75 (0.06)
2	0.07 (0.06)	0.78 (0.06)
3	0.02 (0.03)	0.43 (0.06)
4	0.05 (0.05)	0.71 (0.06)
$\hat{\eta}_j$	0.41 (0.06)	0.59 (0.06)

Members of the first class have small estimated probabilities of answering items correctly. This class is clearly the "non-master" one. Members in the second class have for all items much higher probabilities of answering correctly. This class is the "master" class.

You could re-analyze the data using a latent trait model and compare the expected frequencies obtained under the latent class model and the latent trait model.

Women's mobility

The eight items on women's mobility analyzed in Chapter 7 can be analyzed using latent class analysis. The items indicate whether a woman living in rural Bangladesh could engage in various activities alone.

The aim is to identify groups of women with similar patterns of mobility. You should be able to show that the two- and three-class models give poor fits to the second and third order margins. For the two-class model, chi-squared residuals greater than 100 occurred for many pairs and triplets of items. The three-class model improves the fit considerably, but there are still pairs of items with residuals greater than 20. The four-class model improves the fit even further. The percentage of G^2 explained increases from 72 to 93.7% as we go from the two to three classes, and to 96.7% as we increase the number of classes to four. Table 9.12 shows the two-way margins and the (1,1,1) three-way margins for which the chi-squared residuals were greater than 3 under this model.

Table 9.13 shows the estimated conditional probabilities of responding positively to each item given class membership, and the prior probabilities of belonging to each of the four classes. The classes have a clear interpretation and appear in increasing order of social freedom. Women in the first class have the lowest degree of mobility; the only activities in which a reasonable proportion of women in this class can engage are moving within their locality

Table 9.12 *Chi-squared residuals greater than 3 for the second and the (1,1,1) third order margins for the four-class model, women's mobility data*

	Items	$O - E$	$(O - E)^2/E$
(0,1)	2, 6	40.45	7.00
	2, 8	-25.37	3.80
	5, 8	-42.82	4.79
	6, 7	-14.69	3.88
	8, 5	-42.82	7.74
(1,1)	8, 5	42.82	5.25
(1,1,1)	1, 5, 8	45.77	6.05
	2, 5, 8	45.86	6.89
	3, 5, 8	44.70	5.80
	4, 5, 8	44.81	5.88

and talking to a man they do not know (items 1 and 3). Women in the second class have a greater level of social freedom than those in the first class since most of them responded positively to items 1 and 3 and, in addition, more than a quarter reported that they could go outside their village or town, or visit the cinema. It is estimated that the majority (73%) of women fall into one of the first two classes. Women in latent class 3 have still more freedom; most women in this class can engage in the first four activities. Finally, the fourth class contains a small group of women with a very high level of mobility.

Table 9.13 *Estimated conditional probabilities, $\hat{\pi}_{ij}$, and prior probabilities, $\hat{\eta}_j$ for the four-class model, women's mobility data*

Items	$\hat{\pi}_{i1}$	$\hat{\pi}_{i2}$	$\hat{\pi}_{i3}$	$\hat{\pi}_{i4}$
1	0.44	0.98	0.98	0.99
2	0.01	0.27	0.72	0.89
3	0.46	0.86	0.96	0.99
4	0.04	0.26	0.92	0.98
5	0.00	0.00	0.11	0.78
6	0.00	0.03	0.20	0.96
7	0.00	0.00	0.03	0.80
8	0.01	0.00	0.15	0.82
$\hat{\eta}_j$	0.34	0.39	0.21	0.06

Workplace Industrial Relations Survey, WIRS

The WIRS data were analyzed in Chapter 7 with a factor analysis model for binary variables. The same set of six items can be analyzed using latent class analysis.

The six items, given in Chapter 7, measure the amount of consultation that takes place in firms at different levels of the firm structure. Items 1 to 6 cover a range of informal to formal types of consultation. The first two items are less formal practices, and items 3 to 6 are more formal. The factor analysis for binary items model revealed that the two-factor model was a good fit to the data after item 1 is excluded (the least formal item).

The latent class analysis aims to group the firms with respect to the patterns of consultation they are adopting.

The two-class model fitted to the six items is rejected not only by the overall goodness-of-fit measures ($X^2 = 350.28, G^2 = 299.12$ on 21 degrees of freedom) but also by the large chi-squared residuals for some of the two and three-way margins. All the chi-squared residuals with values greater than 3 include item 1. The three-class model is still rejected ($X^2 = 64.89, G^2 = 67.78$ on 14 degrees of freedom). However, you should be able to show that the fit to the two- and three-way margins is very good. The estimated probabilities for this model are given in Table 9.14. Class 1 represents those firms that mainly use informal policies (items 1 and 2). Class 3 includes those firms that use all the methods but not the first informal one. Lastly, firms in Class 2 use all methods including that under item 1 (with lower probabilities than in Class 3 for items 2 to 6). The last row of the table estimates that the majority of the firms (55%) are in Class 1.

Table 9.14 *Estimated conditional probabilities, $\hat{\pi}_{ij}$, and prior probabilities, $\hat{\eta}_j$, for the three-class model, WIRS data*

Items	$\hat{\pi}_{i1}$	$\hat{\pi}_{i2}$	$\hat{\pi}_{i3}$
1	0.21	0.95	0.06
2	0.59	0.27	1.00
3	0.08	0.43	0.68
4	0.14	0.19	0.62
5	0.11	0.53	0.85
6	0.02	0.25	0.37
$\hat{\eta}_j$	0.55	0.26	0.19

Both the latent class analysis and the factor analysis for binary variables had problems fitting item 1 (the most informal one). In Chapter 7, the analysis was repeated without item 1 giving a two-factor model that was a good fit. We suggest that you should repeat the latent class analysis with item 1 omitted.

9.7 Software

The program LATCLASS (Bartholomew and Knott 1999) available on the Web site estimates the parameters of the model, provides estimates of the asymptotic standard errors, and gives the goodness-of-fit measures discussed in the examples. The estimation of the model in the program is based on an E-M algorithm. A number of other software packages such as Mplus (Muthén and Muthén 2000), LEM (Vermunt 1997), LatentGold (Vermunt and Magisdon 2000), and WinLTA (Collins, Flaherty, Hyatt, and Schafer) are available.

9.8 Further reading

Bartholomew, D. J. and Knott, M. (1999). *Latent Variable Models and Factor Analysis*. 2nd edition. London: Arnold.

Heinen, T. (1996). *Latent Class and Discrete Latent Trait Models*. Thousand Oaks: Sage Publications.

Krzanowski, W. J. and Marriott, F. H. C. (1995). *Multivariate Analysis, Part 2: Classification, Covariance Structures, and Repeated measurements*. London: Arnold.

References

Bartholomew, D. J. (1998). Scaling unobservable constructs in social science. *Applied Statistics 47*, 1–13.

Bartholomew, D. J. and Knott, M. (1999). *Latent Variable Models and Factor Analysis* (2nd ed.), Volume 7 of *Kendall Library of Statistics*. London: Arnold.

Bartholomew, D. J. and Leung, S. O. (2002). A goodness-of-fit test for sparse 2^p contingency tables. *British Journal of Mathematical and Statistical Psychology*. In press.

Bartholomew, D. J. and Tzamourani, P. (1999). The goodness-of-fit of latent trait models in attitude measurement. *Sociological Methods and Research 27*, 525–546.

Basilevsky, A. (1994). *Statistical Factor Analysis and Related Methods. Theory and Applications*. Wiley Series in Probability and Mathematical Statistics. New York: John Wiley & Sons.

Bentler, P. M. (1996). *EQS Structural equations program manual*. Encino, CA: Multivariate Software.

Bock, R. D. and Lieberman, M. (1970). Fitting a response model for n dichotomously scored items. *Psychometrika 35*(2), 179–197.

Borg, I. and Groenen, P. (1997). *Modern Multidimensional Scaling, Theory and Applications*. New York: Springer-Verlag.

Brook, L., Taylor, B. and Prior, G. (1991). *British Social Attitudes, 1990, Survey*. London: SCPR.

Carton, A., Swyngedouw, M., Billiet, J. and Beerten, R. (1993). *Source Book of the Voter's Study in Connection with the 1991 General Election*. Katholieke Universiteit Leuven: Sociologisch Onderzoeksinstituut.

CBSI (1998). Central Bureau of Statistics [Indonesia] and State Ministry of Population/National Family Planning Coordinating Board (NFPCB) and Ministry of Health (MOH) and Macro International Inc. (MI), *Indonesia Demographic and Health Survey 1997*. Calverton, Maryland: CBS and MI.

Clausen, S. (1998). *Applied Correspondence Analysis*. Sage.

Collins, L., Flaherty, B., Hyatt, S. and Schafer, J. *WinLTA 2.0, User's Guide*. The Pennsylvania State University.

Conrad, R. (1964). Acoustic confusions in immediate memory. *British Journal of Psychology 55*, 75–84.

de Menezes, L. M. and Bartholomew, D. J. (1996). New developments in latent structure analysis applied to social attitudes. *Journal of the Royal Statistical Society, A 159*, 213–224.

Ehrenberg, A. (1977). Rudiments of numeracy. *Journal of the Royal Statistical Society, Series A 140*, 277–297.

Ekman, G. (1954). Dimensions of color vision. *Journal of Psychology 38*, 467–474.

Fleishman, E. A. and Hempel, W. E. (1954). Changes in factor structure of a complex psychomotor test as a function of practice. *Psychometrika 19*, 239–252.

Gierl, M. and Rogers, W. (1996). A confirmatory factor analysis of the test anxiety inventory using Canadian high school students. *Educational and Psychological Measurement 56*, 315–324.

Haberman, S. (1978). *Analysis of Qualitative Data. Vol. 1: Introductory Topics*. London: Academic Press.

Hand, D. J., Daly, F., Lunn, A. D., McConway, K. J. and Ostrowski, E. (1994). *A Handbook of Small Data Sets*. London: Chapman & Hall.

Heath, A., Jowell, R., Curtice, J., Brand, J. and Mitchell, J. (1993). *British General Election Study, 1992*. Colchester, Essex: The Data Archive. [Computer file].

Huq, N. M. and Cleland, J. (1990). *Bangladesh Fertility Survey, 1989*. Dhaka: National Institute of Population Research and Training (NI-PORT).

Jardine, N. and Sibson, R. (1971). *Mathematical Taxonomy*. New York: Wiley.

Jolliffe, I. (1972). Discarding variables in a principal components analysis. I: artificial data. *Applied Statistics 21*, 160–173.

Jöreskog, K. G. and Moustaki, I. (2001). Factor analysis of ordinal variables: a comparison of three approaches. *Multivariate Behavioral Research 36*, 347–387.

Jöreskog, K. G. and Sörbom, D. (1993). *LISREL 8 User's Reference Guide*. Chicago: Scientific Software International.

Jowell, R., Curtice, J., Park, A., Brook, L. and Thomson, K. (1996). *British Social Attitudes: the 13th Report*. Aldershot: Dartmouth.

Karlheinz, R. and Melich, A. (1992). *Euro-Barometer 38.1: Consumer Protection and Perceptions of Science and Technology*. INRA (Europe), Brussels. [Computer file].

Knott, M., Albanese, M. T. and Galbraith, J. (1990). Scoring attitudes to abortion. *The Statistician 40*, 217–223.

Kruskal, J. (1964). Multidimensional scaling by optimizing goodness-of-fit to a nonmetric hypothesis. *Psychometrika 29*, 1–27.

Kruskal, J. B. and Wish, M. (1994). Multidimensional scaling. In M. S. Lewis-Beck (Ed.), *Basic Measurement*, Volume 4. Sage Publications. International Handbooks of Quantitative Applications in the Social Sciences.

Lawley, D. N. and Maxwell, A. E. (1971). *Factor Analysis as a Statistical Method* (2nd ed.). London: Butterworth.

Macready, G. B. and Dayton, C. M. (1977). The use of probabilistic models in the assessment of mastery. *Journal of Educational Statistics* 2(2), 99–120.

McGrath, K. and Waterton, J. (1986). *British social attitudes, 1983–86 panel survey*. London: SCPR.

Melich, A. (1999). *Eurobarometer 52.1: Modern biotechnology, quality of life, and consumers' access to justice*. INRA (Europe), Brussels. [Computer file].

Morgan, B. (1973). Cluster analyses of two accoustic confusion matrices. *Perception and Psychophysics* 13, 13–24.

Morgan, B. (1981). Three applications of methods of cluster analysis. *The Statistician* 30, 205–223.

Morgan, B. and Ray, A. P. G. (1995). Non-uniqueness and inversions in cluster analysis. *Applied Statistics* 44, 117–134.

Moustaki, I. (2000). A review of exploratory factor analysis for ordinal categorical data. In R. Cudeck, S. Du Toit and D. Sörbom. (Eds.), *Structural equation modeling: present and future*. Scientific Software International.

Moustaki, I. (2001). GENLAT: A computer program for fitting a one- or two- factor latent variable model to categorical, metric and mixed observed items with missing values. Technical report, Statistics Department, London School of Economics and Political Science.

Muthén, B. O. and Muthén, L. (2000). *Mplus: The comprehensive modeling program for applied researchers, user's guide*. Los Angeles, CA.

O'Muircheartaigh, C. and Moustaki, I. (1999). Symmetric pattern models: a latent variable approach to item non-response in attitude scales. *Journal of the Royal Statistical Society, Series A* 162, 177–194.

Peaker, G. F. (1971). The Plowden children four years later. Technical report, National Foundation for Educational Research in England and Wales.

Rasch, G. (1960). *Probabilistic Models for Some Intelligence and Attainment Tests*. Copenhagen: Paedagogiske Institut.

Reif, K. and Marlier, E. (1995). *Eurobarometer 43.1bis: Regional Development, Consumer and Environment Issues*. INRA (Europe), Brussels. [Computer file].

Ridge, J. M. (1974). Three generations. In J. M. Ridge (Ed.), *Mobility in Britain Reconsidered*. Oxford University Press. Oxford Studies in Social Mobility, Working paper 2.

Roaf, M. (1978). A mathematical analysis of the styles of the Persepolis reliefs. In J. Megaw and C. Greenhalgh (Eds.), *Art in Society*. Duckworth.

Roaf, M. (1983). Sculptures and sculptors at Persepolis. *IRAN, Journal of the British Institute of Persian Studies* 21, 1–159.

Sclove, S. (1987). Application of model-selection criteria to some problems of multivariate analysis. *Psychometrika* 52, 333–343.

Spearman, C. (1904). General intelligence, objectively determined and measured. *American Journal of Psychology* 15, 201–293.

Thissen, D. (1991). *MULTILOG: Multiple, categorical items analysis and test scoring using item response theory.* Chicago: Scientific Software, Inc.

Vermunt, J. K. (1997). *LEM: A general program for the analysis of categorical data.* Tilburg University.

Vermunt, J. K. and Magisdon, J. (2000). *Latent GOLD's User's Guide: Computer Manual.* Boston.

Index